ENERGY WARS
Notes From the Front
Written by William Sargent

TABLE OF CONTENTS

INTRODUCTION
New Bedford

The mayor looked out over his city's depressed downtown then back at New Bedford's optimistic motto, *Lucem Defundo.* " We light the world." It had the ring of an opulent city confident in the knowledge that it was the center of an industry that would continue to light up the world for a thousand years. Whale oil did illuminate the world — but for only 150 years.

In 1850 it cost $2.50 a gallon for premium whale oil, but price didn't matter. Those who could afford it were more than willing to pay for sperm whale oil that provided just the right golden glow they wanted to suffuse their Pre-Civil War homes.

Ship owners built ever larger and more expensive sailing vessels to ply the farthest reaches of the Pacific and Arctic Oceans to bring back the valuable commodity. The industry had already decimated the whale stocks closer to home.

Captains drove thousands of whalers to risk life and limb, so they could build palatial homes and amass glittering fortunes, but such costs didn't matter. Whale oil was king and whaling was the world's most audacious industry with international reach and power. In parts of the world, people thought places like Nantucket and New Bedford must be countries because they saw their names on the transoms of so many whaling vessels.

The whalers reveled in the belief that whale oil was so necessary to civilization that their industry was impregnable — their product so superior, that it would last for a millennium. The Nantucket Inquirer mocked whale oil's rivals as trifling pretenders:

"Great noise is made by many of the newspapers and thousands of traders about lard oil, chemical oil, camphene and half a dozen other luminous humbugs. But let not our envious opponents indulge themselves in such dreams."

But then it was over. The industry had ignored a faux colonel who had figured out how to drill for oil in Titusville Pennsylvania. Edwin Drake's first well produced more oil in an hour than could be found by slaughtering half a dozen whales.

Drake was soon pumping so much oil he had to dump it into empty whiskey barrels and soon discovered he could sell a barrel of rock oil for far less than a cask of whale oil. Today we honor his memory by pricing oil by the barrel even though it comes to us through a hose. Refineries for kerosene lamp oil soon popped up in Boston, New York and throughout the back hills of rural Pennsylvania.

The whalers found themselves in the midst of a transformation that was rapidly changing how the world used energy. Almost overnight one tired, old, worn-out source of fuel was being replaced with another that was far cheaper and easier to produce. The writing was on the wall, but it turned out, the whalers couldn't read. By 1861 the Federal government bought 38 old whaling ships and sunk them in Charleston Harbor in an unsuccessful attempt to blockade the Southern port.

In 2014 I found myself in the faint beginnings of a similar energy transformation. I had naively called a company in East Texas to find out if they planned to frack an old, worn out oil well I had inherited almost by accident. The call started me on this whirlwind exploration of the world of energy as it is, teetering on the edge of a similar energy change.

It took 150 years for whales to recover from the depredations of the whaling industry. We can only hope that it will take 150 years for us to recover from the depredations of the oil industry – to say nothing of what may come after.

Chapter i
An Accidental Oilman
Webster, Texas

During the Sixties my father was running for governor of Massachusetts. He needed to divest himself of energy stocks so he gave my sisters and me part ownership in an oil well he had bought from an old army buddy living in Houston.

Now I know it sounds pretty grand to say you own a Texas oil well, but to tell the truth the well was on its last legs. It had already been in play for several years and its production was down to a trickle. It was only able to pump enough oil to pay each of us about twenty-eight bucks a month. But it was a wonderfully tangible investment. I liked thinking of it faithfully pumping away somewhere down there in East Texas and it allowed me to write environmental books, some of which even bashed big oil. To tell the truth, I liked thinking of myself as J. R. Ewing as well.

But in 1973 I started receiving letters from Exxon explaining that they planned to unitize our oil field. This meant that from then on, we would be paid a percentage of the entire oil field rather than what was being pumped out of our individual well.

But it was a tricky business. They planned to inject pressurized water into the oil field to sweep out the residual oil. It could result in our earning much less than we been making before. It was so risky they offered to buy us out for $5,000 each!

My sisters jumped at the chance. One bought a horse; the other put an addition on her house. But I figured that if Exxon wanted our well so badly, they had to know something this naive Easterner didn't know. So I hung on to my royalties and they plunged to about ten bucks a month.

My Oilwell

I had missed my big chance, but it was still reassuring to know that my faithful little oil well was continuing to provide me with a small but steady income. But it still bothered me to think that maybe I was really more like the poor guy always getting screwed by the Ewings.

In recent months I started receiving almost weekly letters, phone calls and sometimes even already drawn checks offering to buy out my share of the royalties in what they were now calling the Webster Tract. The letters were from small wildcat operators who wanted to use new techniques to extract tertiary oil out my field.

They explained that primary production, just pumping oil out of the ground, had only brought between 30% and 40% of the oil to the surface. Secondary production, injecting water into the unitized oil field, had brought up another 20%, but there was another 20% of the original oil still in the ground that they thought could be brought to the surface using tertiary methods.

It would have been nice if I had just cashed one of those checks for $5,000 no questions asked, but curiosity had gotten the better of me. What did those wily Texans know that this callow Easterner didn't understand? I decided to call up and ask.

I soon realized that, not only couldn't I understand their accents, but I also didn't know what the devil they were talking about. But it did start to sound like they wanted to frack my well!

I have read most of the environmental literature about the effects of fracking, but I have also noticed that the price of gas stayed relatively stable during the crisis in the Ukraine and the Middle East. And I have watched as our neighboring city of Salem Massachusetts decided to replace its dirty coal-fired power plant with a new gas-fired one, thanks to fracking. Plus I also understand the argument that natural gas can potentially be a bridge to get us from an economy based on our dwindling supplies of oil to one based on wind, solar energy and hydrogen gas.

But, did I really think my well was going to lead to energy independence, as some were saying, or just more of the same old environmental problems?

I decided to do some sleuthing. First I clicked on Google Earth to locate the Webster Tract. I found it about 20 miles southeast of Houston. It consisted of about a dozen oil wells just north of Webster, Texas a town of about 3,000 people.

I wondered what all those people thought about having oil wells near their homes; wells that might contaminate their drinking water, but wells that might also be worth millions of dollars. So I checked online.

I discovered that Texas has a hundred year history of having oil wells right beside homes and farms. Cities like Fort Worth, Dallas and Houston boomed because of the black gold often sitting in oily formations directly below their city streets.

I wondered what it felt like to know exactly which formations of limestone; shale or Carboniferous sandstone lay underneath your house. We Easterners never think about such things.

I expected that I would find unconditioned support for oil and gas extraction and people eager to buy into the latest oil boom. Instead I found that Texas was going through about the same debate as the rest of the country.

A North Texas homeowner had just won a $3 million dollar lawsuit against one of the early frackers in April, and the city of Denton would have a referendum on its November ballot to ban fracking altogether. But most of the articles were more nuanced, reflecting the state's long history with hydraulic injection and horizontal drilling and even wind energy. Texas has more wind turbines than any other state, including California.

Mitchell Energy was the first company to drill horizontally to reach the Barnett shale underneath the streets of Fort Worth. Their engineers also went against conventional wisdom and used a slickwater frack that lowered the cost of completing their wells by $75,000 to $100,000. Most importantly, they discovered that by combining the two technologies they could dramatically increase the amount of gas that could be recovered from each well.

When the price of natural gas rose above the equivalent of $90 a barrel for oil in the late 1990's, the two new technologies took off. That initiated what critics started calling the fracking boom. Today, the Newark East Field underlying the Barnett formation continues to be the largest producer in Texas accounting for 30% of all the natural gas produced in the state. Because it is so big and was the first field to be exploited, the early frackers made mistakes and garnered their fair share of environmental critics.

But I also found an online forum for land and royalty owners above the Haynesville shale formation, which covers 9,000 square miles and underlies large portions of East Texas, northwest Louisiana and southwestern Arkansas. It is conservatively estimated to hold 29 trillion cubic feet of natural gas and be worth several billion dollars.

William Sargent
ENERGY WARS

The website demonstrated a new determination to get things right. Its introduction read, "What makes this site so great? Well I think it's the fact that, quite frankly, we all have a lot at stake in this thing they call shale…our farm has been in our family for over 80 years. As exciting as this shale is, we know that we have a responsibility to do this thing correctly. After all we want the farm to remain the place where the family can continue to gather for 80 years and beyond."

But when I started calling around I found that things were even more interesting. Exxon-Mobile had sold the Webster Tract to Denbury Energy in Plano Texas. I called the head of owner relations, Jack Collins, and made my first mistake. I asked him when the company planned to start fracking my well. He replied, by e-mail, "We have no plans to frack Webster Field, but we do plan to commence a CO2 flood of the field next year."

He explained that Denbury owned a source of natural carbon dioxide underneath the Jackson salt dome in Mississippi. They were planning to extend what they called their Green Pipeline to the Webster Field. When it arrived they would start injecting natural carbon dioxide into the well.

The carbon dioxide would adhere to the droplets of oil remaining in the shale. It would be a little like mixing turpentine with paint, the oil droplets would swell and become thinner so they could be pumped up through an adjacent producer pipe.

In 2015 they planned to start piping man-made carbon dioxide from a Mississippi power plant to force more oil out of my field. But after the oil was extracted they planned to sequester the remaining carbon dioxide underground, where it could not cause global warming. The federal government would provide the power plant with a grant to participate in the project.

Jack Collins also sent me some literature that pointed out that the amount of U.S. carbon emissions had dropped in 4 out of the last 7 years due to power plants switching from coal to natural gas. And that a plant in Saskatchewan planned to sequester the same amount of carbon dioxide as would be produced by 500,000 automobiles. It would be like getting 500,000 cars off the road. Had I serendipitously become an investor in an energy company who is doing carbon sequestration and oil extraction correctly?

There had to be a catch. I decided to exercise my journalistic prerogative and fly down to Texas, to snoop around, ask a few impertinent questions and see how my faithful little oil well was doing.

I found the Webster Tract sitting on either side of The Gulf Parkway. The cries of laughing gulls and great-tailed grackles filled the salty air and the silhouettes of vultures circled overhead. A donkey head oil well pumped beside my motel, offshore rigs were moored beside shrimp boats in nearby Galveston Harbor and the Johnson space center was just down the road.

Interestingly enough, the land Denbury owned was some of the least developed along the entire highway. It had large tracts of hardwood forests and the remains of old pear orchards. Horses and dairy cows grazed right beside the capped wells of the oil field. Across the highway towering rigs were starting to drill toward the great dome of salt thousands of feet below my car.

I asked several neighbors what they thought of the project. A carpenter said that he was concerned about reports of earthquakes in West Texas. "Of course there is no way in heaven you can say they were caused by fracking and they really didn't do much damage. But there have never been any earthquakes there before." When I asked a fellow passenger what she would do if a company wanted to frack under her house in Houston, Weezie McKay said, "I would start to look for a new home."

Eric Miller, a chemical engineer in the orchard lands of nearby Orange, said he would be thrilled if someone wanted to use natural gas to get more oil out of his well.

"This kind of thing has been done safely for years. Natural gas is a seasonal fuel used primarily to heat homes in the winter. So when they produce it in the summer they pump it back down into salt domes and store it until the prices rise in the autumn. The chemical industry also gets credits for pumping ethylene into salt domes and of course the government stores strategic supplies of oil in some of the 500 salt domes in this part of Texas and Louisiana."

But perhaps the head of the McIlhenny Tabasco Sauce Company expressed the Zeitgeist best in Mark Kurlansky's book, "Salt", "We are fortunate to live in an area that not only supports agriculture, but has oil, gas and salt as well."

By the end of my trip I discovered that there are many ways of looking at oil. Easterners tend to look at it as a messy business of booming gushers that make Texans instantly rich. But the average well isn't a gusher but one more like mine that can produce a moderate amount of oil for several generations if the correct technology is used.

Scientists see oil as a mineral that built up when our planet was much warmer and plankton was removing heat trapping carbon dioxide from the oceans then sequestering it under tons of sediments where it gradually cooked into oil and natural gas. In essence they see oil as our planet's way of cooling itself down. The problem is that now we are putting that heat trapping gas back into the atmosphere so fast that the natural systems cant absorb it fast enough to prevent global warming.

Thoughtful environmentalists look at oil differently. They see running out of oil as one of our biggest environmental problems. Despite its dirty reputation, oil is one of the cleanest fuels we have. If its price were not continually rising as we eventually run out of this precious resource, the world would not be turning to much dirtier fuels like coal and tar sands and much more dangerous fuels like uranium ore. We could still be converting to wind and solar energy while only using oil for essential things like transportation.

Finally, most Texans still see oil as a pretty good investment that can earn a family a good income for several generations while helping build our nation's economy.

I like to think that using carbon dioxide to extract oil from my well will take all of these considerations into account. My oil field will continue to make a modest amount of money extracting oil out of the ground, without some of the problems associated with fracking for natural gas. It will sequester 4% more heat trapping carbon underground than will be emitted by the pumped out oil.

This 4% differential holds the potential of being able to remove the equivalent of several hundred thousand cars off the road. And the government could require that a larger differential be achieved each year, so that by 2050, 90% of all the carbon dioxide produced was being sequestered. Not bad for an investment that costs a fraction of the price of a Prius.

I think my well also illuminates one of the most important environmental debates of our times. Hundreds of churches, colleges and universities are being asked to divest themselves of their holdings in oil companies.

I support this in principle. Divestiture certainly focuses attention on the problem and makes a great rallying cry but it is a shotgun approach that suffers from one big problem. Energy companies may continue to be pretty good investments for years to come. If well-meaning people divest themselves of energy stocks there will be long lines of less thoughtful investors eager to buy them out. Recently I heard a Boston-based financier boast that he was making a ton of money investing in tar sands, which are the dirtiest fuel of all.

I would prefer to see a more focused approach that would reward companies for working on ways to make money by slowing global warming. Some of these might even be far more effective than those now being envisioned by East Coast environmentalists.

So now I finally had to decide. Would I divest myself of my well? Had I become like one of those big oil conglomerates that were stuck with a stranded asset, an oil well they have paid for but couldn't sell? Clearly not, I knew I could pick up the phone tomorrow and sell my mineral rights to any number of willing buyers.

My well was already owned by a company widely acknowledged to be the leader in the field of carbon dioxide sequestration and oil extraction. Did I think this technology would be the silver bullet to solve all our climate change problems? No.

Did I think there are no problems associated with carbon injection? Also no. The best candidates for CO_2 injection are usually old or abandoned wells that may have not been properly plugged.

In 2011, Denbury paid a $662,500 fine when its injection system blew the cement casing out of an old well in Mississippi and carbon dioxide escaped and settled in the surrounding hollows asphyxiating several deer and several other smaller animals. But even windmills have been blamed for killing bald eagles. Like wind and solar, carbon sequestering is not the complete answer but it does seem to be a step in the right direction.

Unfortunately climate change is inextricably intertwined with two other major problems, our world's inherently fragile economy and the fact that we will inevitably run out of oil. This is what New England engineers call a wicked problem, one so complex and intertwined with other problems that it is "wicked hard" to solve.

Did I think I had stumbled on an investment that could play a small part in solving this wicked problem? Yes I did. Plus I was still curious how this would all play out. So I decided to hang onto my little oil well and see what would happen when we pumped it full of heat trapping carbon dioxide.

Hopefully it would help create a world in which carbon dioxide is declining. If so, I planned to bequeath my little carbon sequestering oil well to my grandchildren for safekeeping.

CHAPTER 2
Unanimous Consent
2008

Congressman Ed Markey couldn't believe what he was hearing. The Chairman of the Committee on Energy Independence and Global Warming had just requested unanimous consent so Bart Stupak could say a few words. The Democratic representative from Michigan was not on the committee but he had spent most of his day patiently waiting to ask some questions.

Calling for unanimous consent is standard practice in such committees. It is how Congress conducts most of its informal business. But Jim Sensbrenner, the ranking Republican representative sitting right beside Markey, said he would not allow the visiting congressman to ask any questions.

Markey was dumbfounded. He liked to run a tight ship and now Sensbrenner was going to block his colleague for no reason other than that he could? The two congressmen went back and forth, until Stupak finally spoke up, quite out of order, warning the Congressman from Wisconsin that the minority party could expect the same kind of treatment from the majority party in future meetings. Sensbrenner merely shrugged his shoulders.

The oil executives smiled. They hated each other's guts and were only too happy to see one of their colleagues called on the carpet. But they shared a special disdain for politicians. John Hofmeister, the president of Shell Oil, remembered being called to testify right after Katrina and

Rita had killed and displaced so many of Shell's workers and their families. It had been an extraordinarily stressful time for the company. It was doing its level best to get its facilities back online, so it could continue supplying gas to the crippled nation.

Then, in the middle of the crisis, Hofmeister had been hauled in front of the Senate and grilled about colluding with other companies to withhold oil to keep prices high! If only the public knew how the energy world really worked. His competitors had been only too happy to steal John's customers while he was tied up with trying to get Royal Dutch Shell back up on its feet.

Oilmen's disdain for politicians stems from how each profession thinks about time. Oilmen have to think decades ahead in order to line up financing, buy mineral rights and conduct expensive seismic work before even drilling an exploratory well, to say nothing of going into actual production.

John was proud of Shell's famous hundred year forecasts that predicted what its experts thought the world's political, economic and energy picture would look like a century in the future. These politicians only thought in terms of tomorrow's headlines and how to get reelected every two damn years. It was the difference between what he called political time and oil time.

To tell the truth, John felt he had more in common with the environmentalists sitting in the back of the room than the politicians in front. At least the tree huggers were thinking about what was best for their grand-children's generation, but the two sides came at the issue from opposite poles.

The oilmen knew that they were going to eventually run out of oil and that it was a dirty and expensive product that needed to be regulated, not because Al Gore was worried about global warming, but because there were a lot of pollutants in oil that were bad for people's health. But they also knew that the world was going to continue to need oil until hydrogen power came online. After all, the world still burned firewood and used whale oil for making watches a century after the former fuels had been displaced by coal and petroleum.

Environmentalists started from the premise that global warming was the problem and if they could just pass a bill to get money out of politics and establish a cap and trade system, the country would switch over almost automatically to renewable energy.

John thought that if Congress just created an Energy Reserve system modeled on the Federal Reserve Banking Program the problems of regulating pollution and running out of oil could all be solved without such political interference. But unfortunately there are no silver bullets to solve these two complex and intertwined problems.

We will need legal and economic inducements plus the combined ingenuity of innovators and entrepreneurs to get the world out of its present tailspin. But few innovators or entrepreneurs were sitting in that room. They were all out in the field like the Denbury engineers trying to perfect techniques to recover the gas and oil that the big oil companies had left in the ground.

Soon both sides would be battling those scrappy little frackers; oilmen because they were showing up big oil and environmentalists because they had just started realizing natural gas was not going to be the clean bridge to the future they had originally thought it would be.

By the end of Markey's hearings it was clear that bipartisanship was dead and Obama's cap and trade bill wasn't going to pass. The president would have to focus his attention on Obamacare and fixing the economy. He still wanted to curb global warming but he would have to wait until he was safely reelected before he dared use his executive powers to solve the energy crisis and curb greenhouse gases instead.

But such gridlock did not stop states from initiating their own energy transformation. That is the genius of our federal system with autonomous states but a national government. So I decided it was best to look around and see what was happening in other states, before venturing back to the international scene.

CHAPTER 3
Baked Alaska
Anchorage, Alaska

2014 was a tough year for big oil. Frackers had exposed the fact that big oil companies were as fossilized as the fuel they were extracting from the earth's interior. While scrappy little frackers could pull together a few hundred thousand dollars to drill into shale in nearby Texas, the big oil companies had to drill expensive wells in the deep sea off Brazil, the jungles of Africa, or the remote regions of the Arctic Ocean. Once called the "Seven Sisters", the big oil companies had become so big and macho they had to keep making major finds just to stay in business.

But once frackers drove the price of oil below $90 a barrel, big oil's investments became stranded assets — assets too expensive to continue drilling for. Plus, if all of big oil's proven reserves were drilled, it would guarantee that the world would become uninhabitable — not a very good long-term business plan.

This was particularly apparent in Alaska. Royal Dutch Shell had paid over $2 billion to buy the exploratory rights to look for oil in the remote regions of the Beaufort and Chukchi Seas, areas also replete with whuloo, walruses, polar bears and other charismatic mega-species.

But on New Year's Eve, Shell's drill ship *Kullak* suspiciously broke free from its tugboat and foundered on the rocks of Kodiak Island. She was declared a total wreck, unfit for further work in the Arctic. This was on top of an accident that had crushed Shell's expensive containment dome. In all she had squandered close to $6 billion, and not found a single drop of oil.

For all her majesty, Royal Dutch Shell looked more like a punch-drunk fighter than the second largest corporation in the world. She had to announce that her 2013 revenues had dropped from $5.6 billion to $2.9 billion, a whopping 49% decline. Then the 9th District Court ruled that the Bureau of Energy Management had grossly overstated the potential value of the Chukchi Sea oil and understated the amount of drill ships, pipelines and infrastructure it would take to develop the field — if they ever found it.

It got so bad that Shell finally had to sell its deep-water assets off both Brazil and Australia in order to concentrate on fracking for gas in the Gulf of Mexico. To top that, they announced they were going to cancel their 2014 drilling season in both the Chukchi and Beaufort Seas.

This meant that 800 Alaskans would be unemployed. But it didn't seem to faze the harbormaster of *Unalaska*, where Shell maintained her drill ships, "We have more than enough business from the commercial fishing fleets."

Alaska's political figures seemed particularly incensed that the Federal Court had backed indigenous people over Shell. The Republican co-chairman of the House Resources Committee Eric Fiege from Chickaloon said rather darkly, "We have had problems with decisions made in the 9th District Court for years. It's not going to change until we change judges."

But in August, Shell announced that it planned to make up for 2014's lost drilling time by using two rigs in 2015. They hoped to have both the *Noble Discoverer* and the *Polar Pioneer* drill in the Chukchi Sea but also have them act as back-ups to each other in case of an accident. Ever since BP's oil spill in the Gulf of Mexico, the Coast Guard and BOEM required that a back-up ship be available to drill a second hole in the event of an accident. Hmm, what could possibly go wrong with having your exploratory vessel also be your relief ship?

CHAPTER 4
Putin's Ploy
Olen Harbor, Norway | July 19, 2014

"Utmerket!" "Molodets!" "Awesome!" Two days after Russian backed rebels shot down a Malaysian airliner over the Eastern Ukraine; Norwegian, Russian and American crew members were congratulating each other as they eased a massive semi-submersible oilrig out of Norway's Olen harbor. The *West Alpha* was on its way to drill for oil in the Kara Sea, between northern Russia and the Novaya Zemlya island chain. But the crew had to hurry, despite global warming; the Kara Sea would only be ice-free for two months.

But there was a big fly in the ointment. After Russia invaded Crimea, Europe and the United States had only slapped Vladimir Putin's wrist. But the downing of the Malaysian aircraft had struck at the heart of Western Europe. Most of the passengers had been from the Netherlands.

Foreign ministers, who had previously been reluctant to poke the Russian bear, were now ready to take action. On July 29th they voted for sanctions aimed directly at Russia's oil industry. If the sanctions were strictly enforced they could both stop *West Alpha's* $3 billion exploratory project and thwart Russia's further ambitions to expand its oil operations into the Arctic and West Siberia's rock shale formations.

This was hitting Vladimir Putin where it hurt. Russia depended on petroleum for 60% of its foreign exports. It was only rivaled by Saudi Arabia as the world's top producer of oil, and after the United States, it was the second largest producer of natural gas, due to the United States recent innovations with fracking. It was a whole new world from what it had been just six short years before.

Putin's Ploy

Russia's position was seriously threatened because its land-based oil fields in Siberia were in serious decline. Unlike Saudi Arabia, Russia had always depended on Western technology to exploit its oil. This was particularly so now.

Russia's largest oil company, Rosneft, had teamed up with Exxon Mobile of the United States and Sea Drill from Norway to drill in the Kara Sea. But Rostneft and its president Igor Sechin were on the sanctions list.

The Russian oil industry also needed to partner with companies like BP, Halliburton, and Royal Dutch Shell to search for more offshore oil and to have access to modern fracking equipment to exploit natural gas in Siberia.

If strictly enforced the sanctions could make it illegal for Sea Drill and Exxon Mobile to provide equipment to explore the Arctic and illegal for Halliburton to supply fracking technology to exploit Siberia's rock shale formations.

But the sanctions had ramifications far beyond Russia's geopolitical gambits. They stood to threaten the world's petroleum supply as well. They might provide the U.S. oil industry with a few more years to exploit its gas reserves without serious competition, but they would make the U.S. and the world even more dependent on Saudi Arabian and other OPEC countries' oil.

But who knew, the sanctions could also spur on companies to develop cars and trains run on electricity – electricity generated from renewable energy sources; like wind, solar and none heat-trapping hydrogen gas.

CHAPTER 5
Oil and Ebola Lagos,
Liberia October 22, 2014

The Liberian President, Ellen Johnson Sirleaf, reread the letter from Rex Tillerson. It was not good news. Since 2006, the Harvard trained banker had been encouraging foreign companies to invest in Liberia. Every year the country exported $930 million worth of rubber, iron ore, diamonds, timber, coffee and other agricultural products; but Exxon-Mobile had always been Ellen's biggest coup, and now Tillerson was saying that Exxon was going to delay drilling Block 13 because of Ebola.

The world was gripped by fear of this gruesome disease. In mid-October the World Health Organization reported that 4,000 people had died from the virus but the real number might be closer to 20,000. They expected that in a few months time, as many as 10,000 people would be catching the fever every week. Modern medicine had made impressive strides but could it still be overwhelmed by this simple non-nucleated virus that caused its sufferers to bleed from every orifice in their bodies?

World leaders realized that millions of people could die in *West Africa* alone, and that large swaths of the entire continent were at risk. America's fragile economic recovery had also tanked temporarily because of the media stoked hype about the frightening new contagion.

Health officials insisted that you could only contract Ebola from someone who had already developed symptoms, so it was much easier to identify than something like the flu. You also couldn't catch the disease through airborne contagion. You could only contract Ebola by coming in contact with the body fluids of a victim, so it should also be relatively easy to contain.

But the media kept pointing to the two Dallas nurses who had worn protective gear and still came down with the disease after caring for Eric Thomas Duncan, the first person to contract Ebola in the United States.

Health workers' bigger fear was that Ebola would spread to Asia and evolve with other viruses, through what is called viral sex, to become an airborne illness, or a disease whose victims became contagious before they exhibit symptoms. This would make Ebola almost impossible to identify and contain. It had happened once before among rhesus monkeys in an experimental lab in Reston Virginia. This was the doomsday scenario that kept health workers up all night with the cold sweats.

It was clear that this epidemic might not be like the polio in the Fifties or the Spanish flu that killed more people than World War I. This could be more like the 14th century plague that decimated a third of Europe's population, more than 75 million people in one fell swoop. Demographers sometimes refer to such epidemics as "slate cleaners" because they can wipe the slate clean of such a large percentage of the world's population.

But Exxon's problems were considerably more mundane. They couldn't find enough workers to man their offshore rigs, to say nothing of their maintenance crews and offices in downtown Monrovia. It was even difficult to find food in the city because so many farmers had died from Ebola in the surrounding countryside.

Exxon had contributed $225,000 to relief groups fighting Ebola and was a long time supporter of an East Texas based charity that operated the hospital ship *Africa Mercy*. But the ship had open wards and wasn't able to deal with such a highly contagious virus. If an Ebola patient came aboard she could potentially infect 400 patients, along with the ship's medical staff and crew from 40 different nations.

For years the *Africa Mercy* had docked in Monrovia, but this year Dave Stevens decided they would send her to Madagascar on the other side of the continent instead, "No port in the world would take our ship if we had an Ebola patient aboard."

One surgeon who trained aboard the *Africa Mercy* had already been murdered by villagers who thought that the cleaning bleach he was distributing to kill the virus was actually the virus itself.

In late October Nigeria announced that it had contained the outbreak of Ebola that had infected 19 people who had been infected by a single traveler from Liberia. To do this they used GPS to track down and isolate over a hundred people who had come in contact with the traveler, but the country was now disease free. A few days later, officials announced that the family, who had lived with Eric Duncan when he had the disease in Dallas, had also emerged after their 21-day isolation period, virus free.

The two cases made it clear that wealthy countries with a well-coordinated health care system could identify and contain the disease, but that poor nations like Sierra Leone, Zaire and Liberia stood to be devastated. Plus it would be the countries' valuable few health care workers who would be most vulnerable.

It was all so sadly ironic. For years drilling companies had encouraged their crews to eat bush meat when working in remote areas where it was expensive to ship in food. This was the way that diseases like AIDS, Ebola, and Marburg Fever had jumped from being diseases found primarily in animals like wild monkeys and fruit bats to diseases found primarily in humans.

It was equally ironic that it had taken an illness like Ebola to stop Exxon from drilling oil off Liberia. Environmentalists had been trying to do this for years, pointing out that every year Exxon spilled the equivalent of an Exxon Valdez worth of oil every year in Liberia, and they had been doing it for over 50 years.

CHAPTER 6
Cooked Alaska
Anchorage, Alaska | October 7, 2014

Dave Bundy pulled out of the driveway from what Jean liked to call their "Old Oil Republican Neighborhood" outside downtown Anchorage. Their house was basically a large cabin with showers and toilets, just not necessarily all in the same room. But since they owned it outright nobody really cared that the house wasn't finished, nobody except the parents of their third son who always stayed in a hotel because they found the Bundy household a little too hippylike for their Mid-Western standards.

Jean waved goodbye to what she called the bankruptcy king of Alaska, then turned her attention back to her graduate studies. She could already feel the mountains closing in for another long, cold Alaskan winter. Winters had been improved somewhat by global warming, but they could still drive you nuts. Winter was the time when the rates of drug use, rape and murder all soared. As Dave always said, most people came to Alaska to make their fortunes, others to escape their ill fortune.

Dave and Jean came to Alaska to get away from the oppressiveness they had both felt growing up in Boston. Dave was the scion of New England's well-known Bundy family whose members had been prominent figures during the Kennedy years. But Dave's had been an austere childhood. He had been 17 years younger than his closest sibling and grew up in a chilly house without other children or animals, so he had turned his energy toward his studies, excelling at Yale then Harvard Law School.

About the first thing Dave told Jean when he met her at a fancy North Shore country club was that he wanted to go somewhere where nobody knew his name. That seemed pretty exotic to Jean. First they tried California, but a chance interview with a partner in an Alaskan law firm sealed their fate. They packed their dog and young baby into the back of their old used car and drove to Anchorage, Alaska.

They soon learned that oil was the opium of the state, the mother who provided everyone a yearly check just for making it through another winter. Most of Dave's clients either needed the checks for themselves or needed their customers to have them — as did all the local drug dealers. The check for 2013 totaled $1,850. Most people spent theirs on luxuries and items of instant gratification. But Dave and Jean always added theirs to their five kids' college funds. It had always helped.

During those early years the Bundies saw how oil permeated everything in Alaska. Everyone supported oil development whether they were liberals or tea baggers. And every third person you met on the street worked for the oil industry, the other two worked for the Federal government or were in fishing.

The yearly checks came from the Permanent Fund established to win support for the trans-Alaska pipeline, which ran from the North Slope to Valdez. It had been largely built and maintained by VECO, a company started by their neighbor Bill Allen.

They had watched VECO grow from a minor oil pipeline service and construction company to a major player in the worldwide oil industry. Then, when the *Exxon Valdez* ran aground on Bligh Reef, VECO had hired 2,500 workers to clean up what was then the largest oil spill in U.S. history. Everyone loved the company in those days, but in 2006, the FBI raided the offices of several prominent members of the Alaskan legislature who had been receiving illegal gifts from the company. It was discovered that over 10% of Sarah Palin's funds for her Lieutenant Governor's race had come from VECO executives.

Finally in 2007 three of VECO's top executives pleaded guilty to charges of extortion, bribery and conspiracy to impede an Internal Revenue Service investigation. Dave and Jean had steered clear of the mess that had ensnared so many of their friends and neighbors, but they had witnessed just how corrosive oil could be when it made up the largest sector of the local economy. It had been like witnessing corruption in Texas in the Fifties or politics in the Middle East, Venezuela and Nigeria, right now.

Horizontal Drilling

CHAPTER 7
Stop Demonizing Natural Gas
Yahlin Peninsula, Siberia

In July 2014 a helicopter flying over the Yamal Peninsula spotted a perfectly round hole in the Siberian taiga. The Internet lit up with con-jecture. Was this proof that the permafrost was releasing heat-trapping methane into the atmosphere or was the hole just a meteor impact crater?

Most scientists agreed that the slick sides of the hole and the lack of partially melted rocks cast out of the hole proved that it had been caused by a methane bubble exploding up out of the earth, not by a comet smashing down through the tundra.

But the arguments missed the true significance of the event. It was yet more proof that there are massive amounts of frozen methane hydrates under the sea-floor and huge methane fields under every continent on the planet.

It was a bubble of methane that blew up the *Deepwater Horizon* oilrig leading to the BP oil spill. Both Russia and the United States went on high alert several years ago when they thought that either India or Pakistan had exploded a nuclear missile in the India Ocean. Scientists finally determined that a frozen chunk of methane hydrate had probably dislodged from the ocean bottom and exploded in the massive fireball detectable by underwater listening devices and overhead satellites.

The clinching argument for many anti-frackers is the visual of a Pennsylvania farmer turning on his water tap and igniting natural gas, the commercial name for methane. What is seldom mentioned is that everyone in that neighborhood remembers how much fun it was to light the methane that naturally flared out of the spigots in the girl's high school restroom years before anyone thought of fracking the Marcellus shale.

So, if we are ever going to solve our climate change problem we need to take another look at natural gas. It is about time too.

Natural gas was the world's first fossil fuel. In 252 B.C. the provincial governor of Sichuan, Li Bing, discovered how to drill wells to retrieve brine for making salt. The wells were as deep as 300 feet. But the drillers soon discovered that evil spirits emanated from the wells were causing explosions and making workers sick.

But, by 100 A.D. the drillers found out that this invisible substance would burn and they had started using it to heat the brine. By the Middle Ages they had perfected all the basic techniques of percussion drilling still used today to drill for oil and gas.

The Sichuan drillers would place a bamboo tube down n hole they had dug, then repeatedly drop a heavy 8 foot long rod with a sharp metal bit down through the tube. This would allow them to go several hundred feet deep, to the gas and brine reserves below. When the gas rose to the surface, drillers would divert it through more mud-lined bamboo tubes to the distant salt works. The Sichuan countryside was soon covered with a spidery web of bamboo pipes used for both home plumbing and transporting gas. The governor also started taxing the operation to expand his empire. Sound familiar?

The Sichuan example points to the extreme versatility of natural gas. You can use it as it is, right out of the ground. It doesn't have to be refined like oil. It can also be cooled into liquefied natural gas, separated into propane or simply pumped through pipelines to cities and homes.

Modern day drillers also have a lot of experience storing natural gas because it is a seasonal fuel that was traditionally used in the winter to heat homes. As Eric Miller told me, the gas would be extracted in the summer then pumped back into salt domes for storage until its price rose again in the autumn. Of course this made it a commodities trader's best dream come true.

The other great advantage of natural gas is that it doesn't have to take millions of years to form like oil or coal. The Massachusetts Water Resources Authority uses month old natural gas collected in its eight giant egg-like methane collectors to help power its waste treatment plant and Chinese farmers use week old methane collected from their pigs' manure to cook their meals.

Oil companies are also experimenting with using natural gas to power their oilrigs and transport vehicles so they wont have to ship expensive diesel into remote drilling areas. These measures prevent methane from being flooded into the atmosphere where it can be 30 times more efficient at trapping heat than carbon dioxide.

But from an environmental point of view the greatest advantage of natural gas is that it emits four times less carbon per unit than coal and two times less than oil. This is the reason that U.S. carbon dioxide emissions have dropped for 4 out of the last 7 years. During the same time countries like Germany and England have increased carbon emissions because their cap and trade policies didn't generate enough income to pay for the renewable energy facilities they had promised the world they would build.

So why does natural gas have such a bad reputation? From what its critics call fracking. Fracking uses a lot of water and causes earthquakes. This sounds dangerous but to date none of the earthquakes have caused major damage. But they have certainly sold a lot of earthquake insurance and undoubtedly lowered property values. But I wonder, just how low your property rate can go if you also own mineral rights to a multi-million dollar gas field?

Many of the problems associated with fracking stem from the fact that we are doing it too fast and too furiously. It has been estimated that natural gas adds a billion dollars to the US economy every day. However, a lot of the natural gas presently being produced should stay in the ground, both to save for the future and also until drillers have the time to build pipelines and compressors to transport the fuel.

If modern drillers had continued to treat gas as a valuable commodity like the early Sichuanese, we would not be in the pickle we are in today. Up until recently oil companies have regarded natural gas as nothing but a dangerous nuisance, better to be flared off than exploited.

Just four years ago, the last thing an oil driller wanted to encounter was a gassy well. As we have seen it was a "kick" of natural gas from a gassy well that led to the BP oil spill. Chunks of frozen methane coming up through the wellhead also nearly exploded the funnel and surface vessel being used to remove the leaking oil.

Today, the brightest lights you see at night from the International Space Station are fires above Middle Eastern and Indonesian oil wells. The drillers are flaring off millions of cubic feet of natural gas because it is cheaper to burn it off than to build pipelines to ship it to market.

Even North Dakota is flaring off so many of its 1,500 new wells that they outshine the city of Minneapolis several hundred miles away. The industry has probably flared off close to fifty years worth of natural gas through this profligate practice and emitted millions of tons of carbon dioxide in the process.

Although environmentalists and governments have tried to limit the practice, it is still legal. In North Dakota you can only flare off natural gas during the first year of production, but this is also when most of the gas comes to the surface anyway. Oil companies can also easily get exemptions in following years if they have not yet been able to construct pipelines.

Then there is all the gas that has escaped into the atmosphere through sloppy extraction techniques and leaky pipelines. It is estimated that huge amounts of natural gas leak out of pipelines below old cities like Boston, causing both global warming and dangerous gas explosions.

So if we stop demonizing natural gas, it can be the bridge we need to lead us from the present age of petroleum to a future age of wind, solar and a hundred percent clean hydrogen gas. But to get there we have to treat natural gas like the valuable resource it is. If we start using natural gas wisely and sparingly it can help us get out of the environmental and economic trap we have constructed to catch only ourselves.

CHAPTER 8
The Pipeline
Boston Common | July 30, 2014

On July 30, 2014, a small group of people gathered on Boston Common to protest the construction of a $6 billion pipeline that would supply natural gas to generate electricity for 1.5 million New England homes. It would also allow power plants to continue converting to the gas, which is, as we know, four times cleaner than coal and twice as clean as oil.

Most of the protesters were homeowners who feared the pipeline would lower their property values. But John Hutchinson-Lavin had a different argument. He argued that Kinder Morgan, the company proposing the pipeline, actually planned to export the gas. I guess he thought folks down in Houston are so clever, they had figured out how to make a killing piping gas from Texas to Boston then shipping it by LNG tanker to the Middle East. I wonder if he might want to invest in my little business plan to ship coals to Newcastle? Other protesters were there to stop companies from extracting natural gas in Pennsylvania.

But the protest got me thinking. As a kid I grew up in Dover Massachusetts. You might have heard of it. Boston Globe columnist Mike Barnicle used to chide Dover for its mile-long driveways and having more trees than people. It was not known for its low land values.

When I was about eight, Algonquin Power laid a natural gas pipeline behind our house. I imagine everyone was compensated for the use of his or her land. I just remember walking our dog through the woods to watch the huge excavators connect and bury the sections of pipes. I must admit I was also intrigued with the idea of the pipeline snaking its way all the way back to the Gulf of Mexico.

But after a few weeks, the excavators disappeared and everything went back to normal. The scar over the pipeline gradually healed to become a nice broad path that now weaves its way through the thick woods. The only problem that I have been able to unearth about the pipeline is that several years ago someone found some invasive plants growing on the woodsy pathway.

Other than that, the pipeline was so well forgotten that neither Dover's present town clerk nor its conservation agent even knew that the pipeline ever existed. Presumably all this time the pipeline has been transporting millions of cubic feet of gas without incident. So much so, that Massachusetts now uses natural gas to produce 60% of its electricity.

This seems to accentuate the point that pipelines have been traditionally considered to be the safest way to transport oil and natural gas. In Houston, where I went to write about my oil well, almost every intersection had markers showing where pipelines crisscross the state.

But the Boston Common protest was not the only anti natural gas action in town. In nearby Salem organizers were planning to hold a "Festival for the Future." They wanted to stop Salem's Footprint Power from converting to natural gas and force it to replace Salem's old coal plant with wind fields instead.

But nobody had any plans to build a nearby offshore wind farm and a proposal to build wind turbines on Salem's Winter Island had been blocked for almost a decade. In the meantime, people had become rather used to flicking a switch and having their lights turn on.

As much as we might want to change over to renewable energy as quickly as possible, we have to face the reality that it cannot be done overnight.

If we are going to make the transition from an economy based on petroleum to one based on renewable energy we have to stop demonizing natural gas. We are going to need gas-fired power plants that can be designed to include renewable energy as those facilities come online. The Salem Alliance for the Future (SAFE) has written that natural gas is "an appropriate if not ideal" bridge to get us to a future based on wind, solar and non heat-trapping hydrogen energy. I would have to agree.

CHAPTER 9

The Voyage of the Whale Watch Vessel Cetacea
Stellwagen Bank | July 28, 2014

"You're Joking Right? That Baby's Going Straight onto E-bay."
- Passenger on being presented a voucher for a complimentary whale watch cruise

On July 28, 2014 the whale-watching vessel *Cetacea* was on Stellwagen Bank looking for humpback whales. It had been a long frustrating day. Too many boats, too much weather, not enough whales. The captain had spent most of his time jockeying for position between fleets of whale watching boats from Provincetown, Plymouth, Boston and Gloucester.

One of the naturalists signaled to the wheelhouse. A mother and calf were off the starboard bow. They had just broken free from the main pod and were swimming west. Perhaps the boats had spooked her, because she was slapping her tail in annoyance.

This was the captain's chance. If he followed the pair he could get them alone and give his passengers one last close-up view before steaming back to the dock. The weather had been dicey all day. A tornado had torn the roof off of 65 houses in nearby Revere, an almost unknown occurrence in New England, and Boston had been flooded all morning. Thick dark clouds were still rolling across the far horizon and the National Weather Service had extended their severe weather warning to 8:00 PM.

The first mate took another look at the chart. They had drifted out of the Stellwagen Bank Sanctuary and were almost a mile inside the Coast Guard's restricted zone.

"That's it. We've got to get out of here!" The captain gunned the engine. If they hurried they could just make it back before their 4:30 deadline. But suddenly there was a heavy thud and the *Cetacea* came to a rattling stop. Perhaps nobody had been standing in the bow looking for lobster traps. If they had, they would have undoubtedly seen the large, bright yellow buoyed cable drifting off the Excelerate Energy's LNG offshore unloading facility.

A crew-member tried to duct tape two ship hooks together to snag the cable but he was comically unsuccessful. The passengers started to grumble. The crew was young and inexperienced and it didn't seem like they had been trained for such an emergency.

The Captain was dead in the water, with 160 passengers in a restricted zone. He sure as hell didn't want to call the Coast Guard, but as the minutes passed it was clear he would have to do something. He radioed his employer, Boston Harbor Cruises, for advice.

They could send two of their other boats to offload the passengers, but someone would have to call the Coast Guard and hire a dive crew to untangle the propeller.

The Coast Guardsmen were not pleased when they arrived. They refused to give the captain permission to transfer his passengers to another boat. The seas were still too rough and the wind too strong. But they did send a doctor aboard to determine if any of the passengers had serious medical conditions. Then they decided that two Coast Guard cutters would stand by the *Cetacea* all night long but the passengers would have to sleep in the rolling boat with nothing to eat but snack food.

Things started to turn ugly when the captain finally announced that the rescue plan had failed. The ship's naturalists had worked tirelessly to pass on the information as they had learned it, but passengers were dismayed that they hadn't heard directly from the captain for five long hours.

The passengers made themselves as comfortable as possible, sprawling across deck chairs and huddling together in bunches on the floor. The best seats in the house were right beside the twin smokestacks. They gave off a little extra heat if you could stand the diesel fumes. The worst part of the long sleepless night was hearing your fellow passengers throw up beside you.

At 3:00 AM a cheer jumped from deck to deck. Divers had finally managed to free the cable and the *Cetacea* started to limp home under the limited power of its single operational propeller. By 8:00 AM passengers were filing down the gangplank to strains of the theme from Gilligan's Island — especially the part about only expecting a three-hour cruise.

The passengers were handed a $50 refund, $500 in cash and a $100 voucher for future cruises. One passenger expressed everyone's feelings, "You're joking right. That baby's going straight onto e-bay!"

What was most fascinating about the whole semi-comedic story was that hardly anybody knew that the offshore LNG facility even existed. What's more, no deliveries had been made to the $350 million facility since it opened in 2008.

The market had changed that much in just six short years. Massachusetts was still getting 30% of its heating fuel from natural gas but now most of it was being produced domestically through fracking rather than coming from the Middle East. This radical change had happened seemingly overnight. I decided it was time to take another look at Boston's LNG facilities.

In 2008 Mayor Menino decided it was too dangerous to have LNG tankers cruise by only 50 feet from the Boston shoreline. And it was far too bothersome to close down the busy Tobin Bridge every time an LNG tanker came to town. After 9/11, Boston had adhered to the Homeland Security's policy but it had never sat well with Boston drivers. "Terrorist smerrorists I've got to be sitting at my desk by nine o'clock!"

But after a Yemeni trained terrorist tried to blow up a Christmas Day flight from Amsterdam to Detroit, Mayor Menino refused to let a scheduled LNG tanker from Yemeni enter Boston Harbor. He was damned if he would let a boat full of flammable gas cruise near Boston's downtown buildings. The beloved mayor eventually won and Excelerate Energy had to build its unloading facility 15 miles off Nahant.

That same year a fully loaded LNG tanker lost power and was drifting toward Cape Cod. It took several days to round up the vessel and tow it into Gloucester where its cargo could be offloaded at an onshore facility. Other than that the facility had lain idle since fracking had gone into full throttle in 2008. It was a testament to how quickly we can expect things to change in a warming world running out of petroleum products.

CHAPTER 10
Nukes and Sea Level Rise
Seabrook, NH | August 5, 2014

The view from the NextEra Energy's boardwalk is sobering. Their Seabrook Nuclear Power Plant sits right behind a beautiful but low salt marsh. Spokesmen for the plant like to point out that the plant is 21 feet above sea level and that a rock revetment and the marsh will protect the plant from being inundated. But they don't like to talk about what is happening underfoot.

For the past decade, groundwater has been infiltrating the walls of the plant's electrical tunnel causing the concrete to lose 22% of its strength due to an alkali-silica interaction. The groundwater floats on a tongue of underground seawater that rises and falls on high tides and with storms. That groundwater lens is also rising at the rate of 6 inches every twenty years due to sea level rise. This is the same mechanism that causes nuisance flooding in cities like Newburyport, Miami and Norfolk, Virginia.

The Nuclear Regulatory Commission was concerned enough about the problem that they had delayed the station's application for an extension of their license renewal from July to October. This was particularly important because Seabrook has applied to extend its operating license from 2030 to 2050.

Senator Markey and Congressman Tierney countered by filing a bill to prevent a nuclear power plant from receiving a 20-year extension if they applied more than 10 year's before their existing license expires. It would seem to make sense to see if you can't fix problems relating to your first 20 years of operation, before applying to operate for a date 16 years in the future.

The last time I was at Seabrook I was camped right beside "Dykes on Bikes" and 2000 other protesters, it was 1977 after all. 1400 of the protesters were arrested and had to spend the night in jail. I had caught the bus back to Cambridge, not a profile in courage, I fear.

But the protest had worked well enough to drive the former owner into bankruptcy and cause the station's second reactor to be canceled. Massachusetts had done its part by blocking the construction of the reactor until the owners came up with an evacuation plan for four Massachusetts towns within 10 miles of the plant, plus Boston and the 4.3 million people who had lived within 50 miles of the plant.

In the case of a severe accident, the entire 50 square mile area would have to be abandoned for 10,000 years because of soil and ground-water contamination. This happened at Chernobyl and Fukushima at a similar plant built by General Electric.

Tokyo Energy Company didn't plan for a tsunami. Let's hope Next Era Energy is planning for more Northeasters, hurricanes and sea level rise, which are all happening faster and faster in New England.

Wind Turbine in Ipswich, MA

CHAPTER 11
Wind Turbines
Ipswich, Massachusetts

When I go to start my car, switch on my lights, or turn up the thermostat I expect something to happen. But, like most New Englanders, I have no idea where all that oil, gas and electricity comes from. I'm fearful of nuclear energy, don't like carbon dioxide, ditto for dams, fracking and oil spills.

But we have very few sources of energy in New England; so if Pennsylvania, Texas or the Middle East want to squander their valuable resources and ship cheap fuel our way, who am I to complain?

We also don't have any derricks or drill ships to remind us where our fuel comes from; just those late night trucks that sneak into town like Santa Claus to fill up our local gas stations.

But things are starting to change. A few neighbors have installed solar panels and when I look out my back window I can see the blades of two wind turbines rotating slowly in the early morning wind.

In mid-August I decided to walk down to Smith Island to take a closer look. The marsh was in all its late summer glory. Last night's high course tide had bent the weak stems of the salt hay marsh grass so now it lay in huge cowlicks. I felt like I was walking on the back of a giant moose. It would continue to grow that way well into autumn.

The full moon tide also drowned the last of the greenhead flies. Unfortunately they had already laid their eggs. They live all year as inch-long larvae that ooze through the marsh mud until they encounter prey. Then they shoot a spiny proboscis into their hapless victim that writhes around in agony for several minutes as its insides are slowly sucked out through the proboscis. The only good thing about these encounters is that the greenheads usually dispatch one of their own relatives.

There is nothing so pleasing to an Ipswichite than to slap and kill a greenhead fly. I say slap and kill because the tough old Tabanids have a disconcerting habit of coming to, and flying away a few seconds after they have been slapped. We have long discussions about technique. Is it better to slap and roll, slap and squash, simply bury them in the sand or better yet drown them in the ocean? I used to bring their little corpses home from the beach to feed our pet turtle.

Of course none of these techniques will make the slightest dent in the future population of the flies. Days before we get around to giving that satisfactory slap, the perpetrator has already laid her eggs from the protein she sucked out of her larval kin. Our blood was just the aperitif to her *plat des jours.* But it is still pleasing to slap this notoriously slow but vicious denizen of the marsh.

I can't help thinking about the greenheads as I walk across the causeway to reach Smith Island. The fieldstone causeway was built around 1740, so the farmhands who lived in the attic of our house could cut the marsh hay without having their horses drown in the marsh.

Building the causeway was backbreaking work. The farmers had to remove heavy stones from the fields, then transport them to the marsh in oxen-drawn carts. Can you imagine hauling tons of boulders in 90-degree heat, surrounded by clouds of deerflies, greenheads and mosquitoes?

When the causeway was completed it stood more than two feet above the marsh, but now the marsh has grown above the causeway. That's because the sea level has risen two and a half feet since 1750. It is a reminder of why the wind turbines are so important. They will not be adding to our atmosphere's surfeit of heat-trapping carbon dioxide.

I'm particularly proud of Ipswich for building its two wind turbines. They now supply 3 percent of the town's total energy and 60 percent of the energy used by our local schools. Unlike many towns our turbines went through the siting process with little fanfare. No residents complained that the turbines would make too much noise, disrupt their television signal or make them ill, because nobody lived very close to the marshy site in the first place.

Perhaps Ipswich should have taken a hint from the town of Hull. She built a wind turbine right beside her high school's football field. The turbine has been credited with scoring several wins for the home team. Their quarterbacks knew to wait until they caught a member of the opposite team looking up at the swirling blades. Then they would snap the ball. It is said Bill Belichick has been seen videotaping their games, and I understand he has asked Robert Kraft to build a turbine near Gilette Stadium.

I'm less proud of the good people on the south shore of Cape Cod. They have blocked the construction of our country's first offshore wind farm for close to 20 years. Their main concern was that the turbines would ruin their oceanfront view, even though the turbines would look less than an inch tall when viewed from the water's edge. It was not lost on the public that many of the same towns that opposed the turbines were not above building replicas of old windmills as magnets to attract summer tourists.

The town of Ipswich and our local high school jointly funded the first turbine. The second was built by a private company, but it sells all its electricity to the town. Together the two turbines supply about 7% of the town's electricity, another 3% comes from the town's part ownership of 10 turbines in the Berkshire Hills. We are well on our way to reaching our goal of getting 20% of our electrical needs from renewable sources by 2020. We may even surpass it.

So Ipswich compares very favorably to Massachusetts, which has the best efficiency rating of any state in the country but only gets 9% of its electricity from local renewable resources. The other 65% comes from shipped in natural gas and 12% comes from coal.

How has Ipswich been able to achieve so much? It had several advantages. First it has owned its own municipal power company since 1903. Though the diesel powered power plant only operated 72 hours in 2013, owning it allows the town to be picky about where it buys its electricity. We now buy electricity from Berkshire Wind, Eagle Creek Hydro and Ipswich Wind as well as from Stony Brook Gas and the Seabrook and Millstone nuclear power plants.

So have there been any problems with the wind turbines? They have already survived two hurricanes, but I understand 6 dead bats were found below the spinning blades of the turbine last summer, circumstantial evidence, but a problem nonetheless. The solution? Town officials turned off the light that was attracting moths and the bats flew off to more productive feeding grounds.

Chapter 12
WIND ENERGY IN THE GRAND MANNER

CHAPTER 12

Wind Energy in the Grand Manner
Sweetwater, Texas

"The answer my friend, is blowin' in the wind. The answer is blowin' in the wind."

-Bob Dylan

If you really want to know which way the wind is blowing, fly down to Sweetwater, Texas. During the late Nineteen Nineties George Bush Senior and Ann Richards set aside their political differences to create the most progressive wind tax credit in the country.

Companies flocked into places like Sweetwater, providing jobs and making farmers and ranchers rich beyond their wildest dreams. And, West Texas became the largest producer of wind energy in the nation. At one point the two richest wind power billionaires in the world both lived side by side each other in Sweetwater and used to josh each other about all their riches during the town's Friday night football games.

Wind energy did so well that people started to look for problems. Mitt Romney led the charge by saying that the federal wind tax program started by George Bush Senior was inherently unfair and that he wanted a level playing field — as if the oil, gas and nuclear industries didn't enjoy grants and subsidies! Grants and subsidies are about as American as apple pie. They are also the reason we have all those nice things we enjoy like universities, High-Tech gadgets, the Internet, and drones.

In 2012 the Koch-funded group, Americans for Prosperity, followed Romney by starting an advertising campaign claiming that "Far left European groups and other radical elements of the environmental movement were behind the tax credit." And Koch-supported represen- tative Mike Pompeo of Kansas sent a letter signed by 52 colleagues to the Chairman of the Ways and Means Committee urging him to let the federal wind tax credit program die when it came up for renewal.

The Texas Tea Party piled on. Their main reason for opposing wind energy seemed to be that President Obama was for it. So they figured it must be some kind of effete East Coast boutiquey sort of energy, not the masculine, polluting kind that Texas should be known for. Republi- can congressmen throughout the country heard the Tea Party message and voted out the federal wind tax credit in 2013. The result? Energy from new wind turbines plummeted by 92% and 30,000 jobs went down the tubes.

Texas Governor Rick Perry had once pledged $10 billion dollars in private investment to the wind industry and was attacked for his perfidy, "Perry joins Enron's Ken Lay and George Bush as fathers of the Great Texas Wind Power Malinvestment."

Of course Perry reversed his position when he ran for President. But the irony remained that 82% of all the wind farms in the United States and almost all the wind farms in Texas are located in Republican Con- gressional districts.

Aside from these political considerations, there are some technical problems with wind energy. They all stem from the fact that wind doesn't blow all the time. This creates additional transmission costs for the $7 billion power lines that connect Dallas and Houston to the West Texas wind fields.

The greatest advantage of wind energy for drought-addled Texas is that wind turbines don't use and pollute massive amounts of water like the oil, gas and nuclear industries. At the present moment, fracking is having its day in the sun, but in the long run, solar, wind and hydrogen gas will be the way to go, and one way to get there will be through tax credits.

CHAPTER 13
"Tis an Ill Wind that Blows no Good."
Nantucket Sound

Cape Wind's Horseshoe Shoals windfarm will be the polar opposite of our Ipswich wind turbines. If it is ever built, it will be the first and largest offshore windfarm in the United States, which is woefully far behind countries like Denmark and Germany.

Those countries have scores of large, elegant, offshore wind turbines that send over 1,000 megawatts of zero emission electricity to towns and cities hundreds of miles away. Germany, Europe's largest industrial power, gets 30% of its energy from renewable sources, which is more than twice the percentage produced by the United States. Denmark gets 40% — embarrassing!

But the United States has benefited from Europe's progress. Their demand for renewable energy lured Chinese and Korean manufacturers into the market, which drove down costs faster than anyone would have imagined when Cape Wind first proposed the Horseshoe Shoals windfarm back in 2001.

The strong winds that blow steadily across Nantucket Sound are Cape Wind's big advantage. They could allow Cape Wind to produce far more electricity than can be produced by land based turbines. Plus, the costs of harnessing that power, if it is built, could continue to drop as the market grows. This is what happened so dramatically in Europe that it put several fossil fuel plants out of business. This is becoming a concern because a few of the traditional plants will still be needed to fill in the gaps when the wind is not blowing.

Cape Wind may be able to supply 75% of the electrical needs of Cape Cod, Nantucket and Martha's Vineyard and experts predict that similar offshore farms could supply a quarter of the energy needed to power the East Coast's many large cities.

So why has it taken so long to get Cape Wind online? Money, power, and politics!

When visionary businessman Jim Gorden first proposed Cape Wind almost twenty years ago, he ran into a gauntlet of rich, well-connected homeowners overlooking Nantucket Sound. They hired lobbyists and lawyers to argue every point and block every initiative. When it comes right down to it, their main concern was seeing the wind turbines rising, what, half and inch above the horizon?

After a few years the issues themselves became unimportant and the debate became what we in Massachusetts call a pissing contest. It didn't make any difference what the arguments were, it just became a matter of who had the most money, time and energy to delay the project.

But Jim Gorden was just as stubborn, and almost as wealthy as his opponents. He spent millions of his own dollars to keep the project on track. And gradually, one by one, Cape Wind passed all the hurdles, and won each of the 20 legal challenges placed in front of it. But even with all the legal cases settled and everything in place to start, Cape Wind's opponents have hired a Harvard Law School professor to argue the constitutionality of one of the minor permitting decisions in one of the already closed cases.

But the project had been formally approved and Cape Wind has an agreement with Siemens of Germany to build the turbines. It has a $600 million loan from the Danish Export Credit Fund and an arrangement to borrow up to $2 billion through the bank of Tokyo-Mitsubishi. Plus it has a $150 million goodwill grant from the U.S. Department of Energy.

Cape Wind already started land construction in 2013 so it could qualify for the Federal tax credit program and offshore construction was due to start in 2015. The windfarm was scheduled to start producing electricity in 2016.

One of the arguments that the homeowners used was that the windfarm would lower their land values. But if they took a trip to Germany's Helgoland, they would see a very different picture. The streets are thronged with tourists and well-paid windmill workers. The workers are staying at a seaside hotel they have rented for the next ten years.

Many of the tourists are on the island to take ecological boat trips to the offshore turbines. One local outfitter, Eike Wallendar told a visiting reporter, "All areas of Helgoland are profiting." We could expect the same thing to happen here, windmills have always been a Cape Cod tourist magnet.

CHAPTER 14
Borrego Solar
Newbury, Massachusetts | August 19, 2014

"You basically have the best policy
environment that there is in the country."

- Joe Harrison, Borrego Solar

On August 19, 2014 Conservation agent Doug Packer closed the
Newbury Conservation Commission's public meeting and the commissioners unanimously approved a plan by Borrego Solar to build 7,000
solar modules on 10 acres of land owned by Ruth Yesair.

The commissioners continued an open meeting on Borrego's second
plan to install 9,000 modules on one of Mrs. Yesair's other properties
zoned for agricultural purposes but near a wetlands area. Together,
the two projects would be able to produce electricity to power 50,000
homes during off-peak hours, and 2,500 during high use times.

It was also not lost on the commissioners that the town of Newbury
stood to collect $50,000 a year if the two projects passed. Perhaps
they could protect themselves by turning down the controversial project
while approving the other one. It seemed worth a try.

Joe Harrison, Borrego's project manager, stressed that Mrs. Yesair was
an environmentalist and felt that this was the best way to protect her
land from traditional development. When the solar lease ran out in 20
years, the land could be restored to its original condition and passed on
to her heirs. That wouldn't happen if she had to sell the land for housing
units.

Joe didn't have to push too hard on the project. Borrego was one of the oldest and largest solar companies in the United States. The company had started in 1980 when astrophysicist Dr. Richards went off-grid by putting solar panels to his home in Borrego Springs, California.

But the company opened an office in Massachusetts in 2007 because the state had the best tax incentive program in the country. The Commonwealth was adding solar generated electricity to the grid almost as fast as Germany. It had doubled its capacity to 462 megawatts in one year, exceeded its goal for 2017, and was already working on producing 1,600 megawatts of solar electricity by 2020. Joe explained to a reporter, "You basically have the best policy environment that there is in the country."

Mrs. Yesair's projects still had to overcome opposition from some of her neighbors. They were concerned that the arrays would lower their property values even though the panels would be hidden behind arborvitae trees.

But the writing was on the wall. It was clear that Massachusetts wanted to build small land-based commercial solar arrays to help provide electricity to its homes and industries without having to import more oil and pollute the atmosphere. It was a small step in the right direction. But was this the right location?

CHAPTER 15
True North
Salisbury, Massachusetts

If you drive up Route 95 from Boston to New Hampshire you can see the largest solar farm in New England. But it has to be a quick look. All you can see from the highway are a few solar panels and the entrance to a hiking path lined with late blooming flowers.

If you are used to visiting nuclear and oil-fired power plants the True North Solar Farm is strangely disconcerting. There is no central office, no workers, no large turbines or generators, just row upon row of panels quietly converting solar power into electricity. You can't even hear the hum of an electric transformer, just the quiet chirping of insects sunning themselves in the grass beneath the panels.

True North was started by a third generation hot water heating manufacturer who owned a plant in the adjacent Industrial Park. When Massachusetts passed the 2008 Green Communities Act to encourage large landowners to invest in renewable energy, Jim Vaughn saw his opportunity.

He had been making more and more solar hot water heaters and realized he was in the perfect position to build a profitable solar farm. He owned 54 acres, already zoned for industrial purposes and had a large roof on his assembly plant that could hold an additional 900 panels to power his shop. His land was south facing, flat, and close to high transmission power lines — the three prerequisites for a successful solar farm.

Although he would have to cut down several acres of trees and create a new wetlands habitat, the ecological benefits of the 6- Megawatt facility would far outweigh the costs. He sailed through the state and local permitting process. Plus he had the added benefit of having no neighbors to complain about property values. The process was easy lifting compared to all the hoops developers of windfarms had to jump through to get their permits.

But financing was more complicated. Jim arranged to have his company, True North, build the facility then sell the underlying land to a Real Estate Investment Trust or REIT, so individuals could invest in the land. Rent from leasing the land would average about $81,000 a year, about as risk free as you can get.

Then True North sold the infrastructure of the farm itself so there would be two revenue streams, one from the land and one from the electricity generated. Revenue from the second stream was higher risk because the price of other fuels used to produce electricity could plummet.

Without long term contracts, the new owners Salisbury Solar could suddenly be under-priced by something like hydroelectric power coming down from Canada or natural gas being fracked in Pennsylvania, even though large-scale wind and solar farms are increasingly producing electricity cheaper than oil and gas-fired plants. This is turning the energy business on its head.

When you think about it, most of the traditional methods used to produce electricity are like using a cannon to kill a fly. You don't need to use radioactive nuclear fuel just to boil water and produce the steam necessary to spin turbines and you don't need to extract huge amounts of expensive oil to do the same thing.

All you need is to have the sun excite a few electrons so they jump from one layer of a photo-voltaic cell to another, thus creating a current of electricity. It all takes place within the thin wafers of silica that make up the panels of a solar farm. It is really just that easy. You don't need huge generators and hundreds of engineers, workers and security personnel.

The only kicker is that the sun doesn't shine during cloudy days or at night. This could be overcome by charging electrical vehicles and car batteries during sunny days and transmitting the electricity back into the grid when it was needed at night.

In the meantime Salisbury Solar continues to supply electricity to the town of Salisbury, the city of Newburyport and Triton High School, the equivalent of powering about 5,000 homes. Plus the communities get a 15% metering credit from the state for using renewable energy. All that is happening right here in cold gray Massachusetts, which is the fourth largest solar state in the nation. Who would have thought?

CHAPTER 16
Hydroelectric Energy
Hudson Bay, Canada

In 1962 I took a 500-mile canoe trip down the Rupert River to Hudson Bay. It literally rained for twenty days and twenty nights. We had to paddle all day in the rain, pause to eat soggy sandwiches, then sleep in wet sleeping bags and get up in the morning to do it all over again. I would have abandoned the trip if the sun hadn't come out just before we came to the only spot on the route where you could flag down a train and go home.

But as the weather improved we started appreciating the beauty of the water soaked muskeg. We ate Canada geese, caught four-foot long pike, and visited several Indian camps where all of us bought husky puppies. Mine cost 25 cents and I had to smuggle her home under my shirt.

The dogs ate, slept, and paddled everyday day right beside us. They became the best-behaved dogs you could imagine. Each canoe prided itself on its own handsome husky jauntily perched in its bow.

One day was critical. We had to have absolutely perfect weather to paddle across Lake Mistassini in one day. It was a hundred miles long and thirty-miles wide. But our luck had improved. The water was like a mirror for the day of our passage.

You could see ripples from our wakes spread out for miles behind us. The gray of the sky perfectly reflected the gray of the water so you couldn't see the horizon and you couldn't see land. It was like being suspended in a big gray sphere of nothingness.

Then all the dogs started howling with fright. It was highly unnerving. Nobody said very much but we all felt vulnerable. The dogs had never done this before.

I just put my head down and concentrated on paddling. We had to make shore before nightfall. I was focusing on the reflection of the sun shining on the water to my right where I was dipping my paddle into the water.

Suddenly I realized something was wrong with the sun. It looked like someone had taken a big apple-sized bite out of its side. Then it started to get very dark and shadows raced across the water. We were in the midst of near total eclipse of the sun.

Today, the muskeg, the Rupert River, and many of those Indian villages have been flooded to supply Southern New England with cheap electricity, if we can just convince New Hampshire to build transmission lines through the White Mountains National Park. But New Englanders don't hold a candle to the cozy little deal Switzerland has worked out with France.

La Grande Dixeme Dam

In 2006 I visited Le Grande Dixeme Dam, hovering in the Alps almost a mile above Lake Geneva. When an American first sees the Alps he thinks of scaling their peaks or schussing their slopes. But when a Swiss national sees the Alps he sees white energy.

Two thirds of all the precipitation that falls on Switzerland gets locked up in what the Swiss like to call their eternal glaciers. When the glaciers melt in the spring they release all that potential energy into bubbling cascades of life giving, life enhancing water. For centuries the Swiss harnessed these glacial waters to irrigate their pastures, meadows and vineyards.

But after World War II, Switzerland realized it could also harvest these waters for energy. The Swiss government built over a hundred miles of underground tunnels to direct melt water from 35 major glaciers in the Valais Canton, and stored it in a mile-long lake behind a massive gravity dam 2,400 meters above sea level.

The dam is probably the least known major engineering structure in the world. It is more massive than the Great Pyramid of Egypt, rises higher than the Eiffel Tower, is longer and was built at a higher elevation than the Hoover Dam. And there I was, an American tourist, thinking nothing was bigger than the Hoover Dam.

In fact if you poured all the concrete used to build the dam into a five-foot wall it would encircle our planet at the equator. But a Chinese journalist in our group pointed out that the Three Mile Gorges Dam would be even larger.

"Ah oui, c'est vrai," said our Swiss host. "But we did not have to displace a single human being. You will displace what? Several hundred million people in many cities and villages, non?"

Le Grande Dixeme has remained unknown even to most Swiss tourists, despite a charming film that hikers can see only after descending deep into the dam's living, breathing, concrete heart. Despite its humble use as a touristy documentary, the film had the distinct feel of a classic. We saw canister after canister of cement being ferried overhead in dizzying aerial shots, and we saw untold tons of concrete being dumped and compacted in preexisting molds.

It was only when the credits rolled that I realized this was the first film ever made by Jean- Luc Goddard. As a student the famed French director poured concrete into the dam by day, and poured his heart into his epic film all night.

After twenty years of round the clock construction, the dam was finally finished in 1961, three years ahead of schedule and way under budget. Today Le Grande Dixeme helps Switzerland produce more than 60% percent of its electricity through such hydroelectric dams. Their hydro-electric power is so plentiful that the Swiss government sells it to France during the day when the rates are high, and then buys back nuclear generated electricity from France at night when the rates are low.

The Swiss then use this cheap French power to pump water back up behind the dams so they can sell more electricity to France the next day – at thirty times more than they paid for it!

Switzerland uses all this clean, cheap hydroelectric energy to run the vast network of electrically powered trains and subways that weave through the countryside like the inner workings of a finely tuned Swiss watch. It is so efficient that after Fukushima the Swiss parliament was able to quietly vote to phase out its five, almost never mentioned, nuclear power plants by 2034.

CHAPTER 17
Geothermal Energy
Bardarbunga, Iceland | July 18, 2014

On July 18, 2014, Bardarbunga, Iceland's largest volcano, started rumbling like an old man's chili filled bowels. It brought back bad memories.

The last time one of Iceland's volcanoes erupted it caused massive floods and the largest closure of European airports since World War II. In 1783 Iceland's Hekla volcano spewed out enough lava to bury Manhattan to the top of the Rockefeller Center. It also killed three quarters of Iceland's livestock, which led to widespread human starvation.

Benjamin Franklin was in Europe at the time, negotiating a treaty to end the war of Independence and wrote a paper that noted that the 1783-84 winter was so severe because ash from the eruption had cooled the planet. It was the first time anyone mentioned the link between climate change and volcanoes, no mean feat for Poor Richard.

But Hekla was nothing compared to the climatological disturbances caused by Indonesia's Tombora volcano that erupted in 1815. It cooled the world's temperature so much that Lord Byron, the poet Percy Shelley and his young paramour decided to stay inside their cabin on Lake Geneva and see who could write the best ghost story. Nobody remembers what the great poets wrote, but the 18 year old Mary Shelley penned *Frankenstein*, perhaps the greatest horror story ever told about the dangers of science unhinged from morality.

Not far away their countryman William Turner was painting "Red Sunset over Lake Geneva." His signature red sunsets did not just spring from his artistic imagination; they were also caused by ash from a volcano on the other side of the world.

New Englanders remembered 1816 as the year without a summer. It snowed during every single month and farmers took to calling the year, "Eighteen hundred and froze to death."

But at least commentators could pronounce Bardarbunga's name. In 2011, nobody could pronounce Eyjafjallajokal until Bill Maher figured out you had to say, "I'm a fallopian tube," with just a bit of a Scandinavian lilt and a lot of Icelandic beer.

There are certainly drawbacks to living on an island where a hot spot of magma has pushed the mid-Atlantic Ridge above the surface of the ocean. But Iceland is also the only place in the world where you can see the European and American plates rifting apart. The only other spot where you can see sea floor spreading on land is alongside rhinos and elephants in The Great Rift Valley of Africa.

It is Iceland's combination of this deep reserve of magma, plus multiple shallow volcanoes caused by rifting and copious reserves of groundwater recharged by 177 inches of precipitation a year that also make the northern island the best place in the world to develop geothermal energy.

The first person to tap into this source of perpetual energy was Snorri Sturlusson, who built a pool heated by geothermal energy in his back yard ... about a thousand years ago. You can still see Snorri's pool, restored in the town of Reykolt. Its waters are just as hot today as when he built the pool in the Middle Ages.

A few years ago I was driving along the Ring Road south of Reykjavik and noticed pillars of steam rising out of the ground and hot water streams coursing down a beautiful green mountainside valley. I drove off the highway to take a closer look. I had stumbled, by chance, into one of the most fascinating communities in the world. Geysers erupted regularly in Hveragerdi's town square and billowing clouds of steam enveloped a guest-house perched on pillars over the hot waters of Hveragerdi River, whose banks are lined with romantic little riverside hot tubs.

In the early days, entrepreneurs used to fill horse drawn carts with loads of dirty laundry in Reykjavik and drive them thirty miles to be washed in the boiling waters of the Hveragerdi river. The cleaned clothes would be returned back to Reykjavik neatly folded the following day.

Other than that, Icelanders didn't use their geothermal energy for much of anything except baking hot spring bread they called Huerabrauth. So for centuries Iceland was one of the poorest nations in Europe, totally dependent on peat and imported coal for energy.

Even as late as 1970 most of Iceland's homes were heated with oil, but today Iceland has one of the highest living standards in the world and it gets 99% of its electricity from renewable sources.

It was the 1970 oil shock that spurred Iceland to do something about developing its geothermal potential. Reykjavik Energy started investing in pumping naturally pressurized potable water up out of the ground from hot magma sources in the surrounding countryside and shipping it through massive pipelines to provide hot water and heat to Reykjavik homes. You still get whiffs of sulfur dioxide when you turn on the tap in downtown Reykjavik. It is the earthy smell of our planet's molten core.

Today, hot water is so abundant and inexpensive in Iceland that even small towns have their own Olympic sized heated pools and swimming, drinking and hanging out in hot tubs seem to be Iceland's national pastimes.

The Hveragerdi area also provides Europeans with over 60% of their winter vegetables. The vegetables are grown in massive hot water heated greenhouses that are built to withstand Category 3 Hurricanes. It is now estimated that if Icelanders were still dependent on oil, their heating costs would be 5 times higher than they are today.

Iceland also uses its geothermal energy to generate electricity. The best place to see this is floating on your back in the turquoise waters of the Blue Lagoon. If you peer carefully through the steam you can just see the towers of the Nesjavollir geothermal power plant towering over the pool.

The plant generates electricity for Reykjavik by using steam from ocean water it pumps underground to be heated by magma. Over the years the plant pumped millions of gallons of marine wastewater filled with the boiled bodies of dead plankton into the surrounding area. Eventually the calcareous shells of all the dead plankton clogged the pores of the porous black basalt to create the Blue Lagoon with its preternatural color and soft skin enhancing waters. It is now Iceland's largest tourist attraction, not bad for an industrial wastewater site.

Electricity is so cheap in Iceland that the nation has become a major center for the aluminum industry that requires large amounts of electricity to smelt bauxite that it ships in from Africa to make aluminum.

Geothermal building at North Shore Community College, Danvers MA

It is no secret that both Microsoft and Google have also looked into whether Iceland could provide the huge amounts of electricity they need to power the computers that run the World Wide Web.

But Iceland is also investigating how to use its abundant supplies of electricity to covert water into hydrogen gas to fuel their automobile and fishing fleets.

Hydrogen powered engines don't pollute because the hydrogen simply bonds with oxygen during combustion to change back into water. In other words, Iceland is pursuing the holy grail of our planet's energy future. Snorri would be proud.

CHAPTER 18
Locally Grown Geothermal
Danvers, Massachusetts

The first thing you notice about the Health Professions building on the Danvers campus of the North Shore Community College is its striking design. An array of raised solar roof panels gives it a soaring aeronautical look.

Inside, two nursing students quiz each other for an upcoming exam in an open, airy sustainability lounge. A quote from the Nobel Prize winning physiologist Albert Szent Gyorgi is etched into a floor length window flanked by bamboo trees and flat screen monitors that display up to the minute readings of how much electricity the solar panels are producing.

Even on a cold, dark day in November this building produces as much energy as it consumes. On sunny days and holidays it produces so much excess electricity that it earns the college credits and cold cash from the town of Danvers.

But it is the things that you don't see that are the most impressive. Sixty 500-foot deep wells and ten miles of pipes lie in closed loops below the adjacent parking lot. They pump water into the ground where it is warmed by the earth's constant interior temperature to 55 degrees Fahrenheit.

The water is then pumped back to the surface where it is either compressed to heat the building in the winter, or piped to inconspicuous chill beams in all the classroom ceilings that cool the building in the summer, without using loud, energy consuming fans.

Natural light softly diffuses throughout the building from narrow rooftop windows that run the length of the building. Their light is directed down through daylight wells to illuminate the corridors and stairwells of all three stories of the building. Large public spaces and classrooms are all situated on the south side of the building where they receive natural daylight through specially angled light shelves built into the building's exterior.

Even the doors of the classrooms are made of sustainable "wood", fashioned from laminated bamboo shoots that grow as fast as a foot every day. A wall panel explains that bamboo grows so fast because it is actually a type of grass and not a traditional woody tree.

The basement contains the pipes, tanks, compressors and Shami Quazi, the engineer who makes the whole thing work. North Shore is the only Massachusetts community college that has its own engineer. He sits at his desk in the basement occasionally glancing at a laptop computer to make sure that the entire system is working properly. He reminds me of a chief engineer on a large ship, but without the dirt, noise and smelly diesel fumes.

How did this all come about? It was the brainchild of Wayne Burton, president of the college, amply supported by Governor Deval Patrick. They were able to raise an extra $2 million dollars, needed to dig 10 extra wells and soaring solar panels that celebrate the fact that this is the first Zero Net Energy Building in Massachusetts and the fourth largest in the country — not bad for the $24 million dollar facility that trains future health workers and sits in our own backyard.

CHAPTER 19
Biomass
Bermuda

Most people go to Bermuda to lie in the sun. I go to visit the island's waste treatment plant. But this is not just any waste treatment plant.

Islands have to be careful about how much energy they use and how much waste they produce. Because Bermuda has one of the highest living standards in the world, it uses a lot of energy and produces a lot of waste. But it also has an extremely strong environmental ethic.

So the island decided to kill two birds with one stone. In the Nineties they closed all their landfills and hired a Swiss firm to build a biomass facility that would burn all the island's combustible trash and use it to produce electricity.

Actually Massachusetts built the first commercially successful biomass facility. It manufactures wood pellets through a company in Burlington. But I decided to take my science writing class to Bermuda. After all it was March... in New England.

Belco, the Bermuda Electric Company, gave us the grand tour. We watched their garbage trucks converge on the plant from all corners of the island. We saw them dump the trash, chop up the wood products and mix them with just the right amount of paper and cardboard, then push the entire mixture down the chute toward the mouth of the waiting incinerator that burns at about 1000 degrees Fahrenheit year-round.

Smoke rises up through an electrostatic scrubber in the smokestack, removing 99 percent of the particulate matter. This is mixed with a slurry of cement and ash that is removed from the bottom of the incinerator. The slurry is then used to make 2 ton blocks of concrete that are made into seawalls. We saw long lines of them fortifying the airport's runways when we arrived on the island.

Finally the incinerator boils water into steam, which is used to spin several turbines to produce enough energy to run the entire plant and still supply the island with 3 percent of its electrical power. That is enough to replace one of Belco's diesel generators that must use costly, imported fuel.

Normally burning biomass to produce electricity is not a very efficient proposition. Because trash is such a low- grade fuel, most of its energy goes up the smokestack along with carbon dioxide and toxic dioxins. But if you are an isolated island in the middle of the Atlantic Ocean that can burn all its combustible waste, make concrete blocks for badly needed seawalls and still supply 3 percent of your electricity needs, it starts to make sense. Plus, Belco is now urging homeowners to install and to build solar panels on their roofs for hot water heating.

To top it off, Bermuda, like Iceland, has made a tourist attraction out its waste treatment plant. At the end of our tour Belco officials urged us to climb to the top of the 200-foot smokestack. It provides the only legal aerial view of the entire island. Helicopters and private drones are banned from Bermuda because they interfere with peoples' privacy.

CHAPTER 20
The Problem is Corn Not Drought
Ames, Iowa

My daughter is a pretty typical recent college graduate. She has hipster friends who work for Facebook and statistics friends who work for investment banks. Like most bicoastal graduates she has been led to believe that fancy young financiers on the East Coast and hip, young techno-wizards on the West Coast are the new masters of the universe. But in 2012 she received a terrible shock. She read that Mexico buys all of the corn it uses to make its iconic tortillas from the United States and that kernels of corn, not derivatives or Facebook ads are actually among our largest export items.

The best way to understand this situation is to visit a typical Mid-Western farm, like the ones that were devastated by the 2012 drought. You would find that most of the farms in the region are dependent on only two crops, corn and soybeans, that are grown in rotation. Productivity is normally phenomenal on such farms. Three farm hands can produce over two million dollars in revenue from a farm the size of seven Central Parks. They do this through the same sort of high technology that has increased the productivity of other sectors of our economy.

They plant and harvest corn in huge GPS guided combines that can cost half a million dollars each. Farmers sit high up in the combines' air-conditioned cabs that look more like the cockpit of a modern fighter jet than the seat of an old fashioned tractor. Monitors tell them how many bushels of corn they are harvesting per acre, how many corn stocks are being left in the field and the moisture content of the soil. Cruise control switches speed up and slow down the combines according to sensors that tell them how many corncobs there are per plant. There is even a refrigerator to keep the driver's beer cold and a buddy seat so the farmer's grandchild can go along for the ride, plus some combines are now driven by remote control.

An East Coast urbanite might be forgiven for thinking that farming from the cab of such a sophisticated piece of machinery is more like watching the Red Sox from the air conditioned confines of an expensive skybox than watching the game from the Fenway bleachers.

The other thing that an East Coast urbanite is likely to miss is that very few of these farmers still plow their cornfields. They simply spray herbicide onto their barren fields in the spring then plant corn seeds that have been genetically modified so they wont be killed by the herbicide that is busily killing the weeds and wild grasses.

Such labor saving practices have allowed farming productivity to soar. Prior to World War II it took a hundred man-hours of labor to produce a hundred bushels of corn, today it takes less than two! Another way to look at that statistic is that the United States now has more bus drivers than farmers, yet we still are the world's top exporter of agricultural products.

The reason for all this productivity is corn. Research, technology and subsidies have made corn one of the most productive crops on the planet. It is now cheaper for most traditional corn growing countries to buy cheap corn from the United States, than to grow their own.

But an East Coast urbanite should not be forgiven for missing that the real problem out here is not so much drought, but our over reliance on corn. Unfortunately, we are going to have many more extreme droughts during the coming thirty years. There is little we can do about it. Even if we could stop the release of all greenhouse gases tomorrow, such extreme droughts would continue because of the atmosphere's backlog of climate warming carbon dioxide. The northern Mid-West already has the climate of Kansas and in twenty years is predicted to have the present day climate of Texas.

But if we can't do much about stopping the amount of global warming that is already in the system, we can do something about the crops we grow. Corn's great weakness is that it is uniquely vulnerable to global warming. As an open pollinator, it does not pollinate itself internally but must rely on the whims of nature to cross-fertilize. This means that corn plants have a crucial seven-day window of opportunity when there must be enough moisture in the air for them to cross-fertilize with other corn plants. If corn does not have these conditions its kernels will sponta-neously abort as they did so disastrously in 2012.

Nature and free markets could solve all these problems, if it were not for artificial constraints. There are a handful of large companies that are making huge amounts of money by turning corn into cattle and poultry food, corn syrup, fructose and ethanol. They employ an army of lobbyists busily working to maintain the taxpayer subsidies and insurance programs, which continue to make corn king on Capitol Hill.

Corn is the problem, not drought

Without these artificial constraints the production of corn would gradually move into Canada as the climate warms, Mid-West farmers would start to grow more drought tolerant crops for food, and non-food crops like switch grass and sugar cane stalks to make ethanol. Taxpayers, the world's food supply, and the environment would all be the beneficiaries of such gradual evolutionary change.

CHAPTER 21
Cows and Chickens
Ipswich, Massachusetts | August 22, 2014

If corn, not drought, is the problem, then what is the answer? It might be right here on the rolling hills of Appleton Farm, the oldest continuously operated family farm in the United States.

You don't see thousands of unhealthy cattle standing knee deep in their own diarrhea because they have been force-fed an endless stream of subsidized corn. You don't see them being castrated and vaccinated with antibiotics and bathed in pesticides to ward off the cattle diseases, which run rampant in the overcrowded feedlots.

Instead, you see about 40 gentle brown Jersey cows and twice as many calves and beef cattle wandering over 500 lush green acres of fertilizer–free pastures. Twice a day the cows saunter down this well-worn path to the milking barn where they provide fifty thousand gallons of milk, cream, butter, cheese, and yogurt to hundreds of people in the surrounding communities.

Any manure the cows leave behind is composted with woodchips from the farm's extensive woodlots and spread on the farm's community supported agricultural fields that produce three hundred thousand pounds of fresh fruit and vegetables to over 800 people who eagerly pay $650 a year for the privilege.

A mobile chicken coop is being moved into one of the pastures that the cows have recently vacated. Their manure patties have matured for a few days so now their rich fauna of grubs and maggots has grown and hatched. The chickens busily scratch apart these partially dried patties to pick out their favorite morsels. By the end of the day the chickens will have devoured enough larvae so the cows will not need to be bathed in pesticides and the chickens will move on to the pasture. In their wake they have left behind a nice, thick mat of nitrogen-rich chicken manure to refertilize the grass for when the cows return in three weeks. In the process they will have recycled and sequestered carbon back into the grasslands and soil.

They will have also produced over a hundred and fifty thousand free-range eggs that will be distributed along with the milk and meat through the farm's food cooperative.

The system recreates an ancient ritual. For eons birds have followed herbivores, to feed on the insects the ungulates stir up in their slow passage across the plains and savannas of the planet. Cattle egrets still follow water buffalo in Africa, black birds still follow bison in America, and pterodactyls once followed Brontosauri during the ancient Cretaceous.

This farm is a closed self-contained natural system. The sun provides the energy to grow the grass. The cows digest the grass to provide protein for us; the carnivores, and the chickens provide the services to keep the cows healthy and free of parasites. There is no need for petroleum to supply outside energy to plow and fertilize the fields — no need for petroleum-enhanced corn to fatten the cattle for slaughter. It is a natural system in which plants and animals do all the work of recycling carbon and nitrogen through the environment, which is constantly being enriched and renourished.

Appleton Farm, Hamilton, MA

It may seem like a revolutionary new idea, but it has been practiced on this same piece of land since a hundred and fifty years before the American Revolution. And it is dependent on a host of grasses that have evolved over the millennia to tolerate changes in the amount rain, wind, drought and heat on our ever-changing globe.

It is a lesson in adaptation well worth remembering as we delve deeper and deeper into this modern age of global warming.

CHAPTER 22
Black Gold
Brick Ends Farm | Hamilton, Massachusetts

One of the miracles of our universe is that you can grow food in dirt. But it is also one of the miracles we take most for granted. What happens after you grow and prepare food and have to get rid of the scraps? In nature they get recycled back into the earth, but in modern America, more than $55 billion worth of food waste ends up in landfills every year.

If you live on the North Shore, much of that waste ends up at Brick Ends farm in South Hamilton. Every week over 100 tons of garbage is delivered to the farm to be recycled back into fine rich compost, "black gold" in the trade; and it really isn't simple dirt after all.

But I wanted to see for myself, so on a stunningly beautiful fall day I drove out to South Hamilton, and found the farm on a winding street lined with cows, barns, horse farms and the bright green playing fields of a private school. The maple trees were in all their autumn glory, hardly a place where you would expect to find a waste management facility.

In fact I had to turn around twice. The only sign marking the entrance to the farm was a small metal plaque, no different than all the other discrete signs on this quiet little road.

The lack of a sign didn't seem to deter anyone else. A steady stream of trucks from Boston Area schools, colleges, and waste disposal companies drove past the silos of dark brown mulch, farm manure and a pond filled with geese.

In the woodsy unloading area, I watched a truck from Gloucester dump a ton of squid offal onto a large pile of food wastes, followed by trucks from a curbside composting program in Methuen, Hannaford's super markets and Boston University's food cafeteria. They were interspersed with smaller pickup trucks dumping leaves, bush and yard clippings onto another huge pile. There were no fences, signs or discernible check-in points, but everyone seemed to know just where to go and what to do.

The first thing I noticed was how clean and neat everything looked. Normally I would have expected to see flocks of seagulls picking through the trash and smell some pretty strong odors. But all I could detect was a slight earthy smell even though it had rained for four days during a recent Northeaster.

I didn't notice a wide mesh network of lines strung high overhead. It was explained to me later that gulls wont fly over, under or through such lines. But there weren't even any gulls lurking outside the perimeter. Amazing! Land rats were kept in check by a pack of local, no doubt overweight, coyotes that made midnight raids on the vermin.

I also didn't see many people. Only four men were running the entire operation. Some were driving front-end loaders to mix the food wastes and leaves together in a 4 to 1 ratio and then shape them into a long high row which would be mixed later with wood ash from neighboring fireplaces.

The only non-digestible elements were plastic bags that stuck out of the pile like fluttering talismans. Every year the crew had to spread out the pile, and pick out all the non-compostable trash bags by hand. Some years they hauled up to 16 tons of plastic bags to the nearest commercial incinerator. Other non-digestibles were easier to digest. Apparently Harvard students were particularly negligent about tossing their silverware in with their food scraps — thinking of loftier things, no doubt. But the crew discovered that there was a pretty good market for such silverware and some of the better pieces even found their way into the owner's kitchen drawer.

The finished pile would sit in the sun for four months while aerobic bacteria broke down the carbon in the leaves steadily generating 160 degree heat, even under thick blankets of New England snow. After four months all of the carbon would have been consumed and only worm castings and nitrogen rich compost would remain.

Another front-end loader was dumping a mixture of this weathered compost into a long green filtering machine. One conveyor belt carried a small stream of large indigestible chunks out of one side of the machine and another carried a much larger stream to the top of a 60- foot high pyramid of dark rich compost. .

Every year, 15,000 cubic yards of this rich black gold will be spread on local farmland and sold to garden centers throughout New England. People from the towns like Hamilton and Ipswich who paid $50 to participate in the their town's curbside composting programs with the farm will get a 50 pound bag of compost free for their efforts. The programs save each town about $30,000 in tipping fees each year that they would normally have to pay if they had the wastes burned in a traditional incinerator.

But I also couldn't find anyone to talk to. Finally a man with a craggy New England smile came sauntering along, walking his none too imposing dog. It turned out he was Peter Britten, owner of the operation.

After teaching in Boston for several years he had moved back to his wife's family farm in Hamilton, which was going through a Perfect Storm of change. Dairy farms were being sold left and right to be developed into Mac-Mansions. Most of the farms that still hung on had switched over to growing "Round-up" corn, which was much easier to grow and more lucrative than raising cows.

But the corn required expensive petroleum based fertilizers that polluted streams with phosphates and created dead spots and algae blooms when rivers finally washed the nitrates into the ocean. And of course the farmers would have totally depleted their soil and would be forced to start the whole cycle over again.

But Peter saw this Perfect Storm as an opportunity to convince farmers to renourish their farmland with compost. Today he has customers throughout Maine, New Hampshire and Massachusetts. It is another one of those small New England miracles, improving our land while transforming how we use our scarce food and energy resources.

Seneca Freylewe and friend

CHAPTER 23
Wits and Water
Ipswich, Massachusetts

August 26 2014 was a beautiful late summer's day. The sun shone through the cobalt blue sky that already held hints of coming cool weather. The marsh was streaked with crimson slashes of Salicornia and dewdrops glistened on clusters of ripening beach plums. Stalks of goldenrod heavy with blossoms swayed in the early morning breeze as I walked down a rutted path that led to the Crane's Beach dock.

I was following some volunteers from the newly formed Ipswich sustainability group that calls itself "Living Room." Four of us strained to lift a large and heavy green crab trap. As soon as it reached the surface, all you could hear were the scuttling sounds of 500 trapped crabs. They sounded like the swarms of scarab beetles on the sound track for Raiders of the Lost Ark.

The first thing a local sustainability group does is identifying what makes their community uniquely sustainable. This is easy for Ipswich. Our clam flats have supported human populations since Paleo-Indians first walked these shores.

During times of want and depression many of us still fall back on this free bounty. All that it requires is a strong back and a small boat. And today, 250 commercial clammers make a good living, and Ipswich clams remain world-renowned.

But in recent years there has been an explosion of green crabs. These voracious predators can crush and break juvenile lobster shells, nip off the necks of steamers and will even start eating marsh grass and eelgrass blades when overpopulation makes the Hunger Games really begin.

Green crabs are now considered to be among the top one hundred most destructive invasive species on our planet. They originally came from Northern Europe but because they can tolerate such a wide range of temperature and salinities they have been able to colonize Africa, South America, both the East and West coasts of North America, and even Australia.

The crabs were first reported in Massachusetts in 1871, but in recent years their numbers have exploded so much the insatiable decapods have become an indigenous species. Nobody is quite sure why their numbers have increased so rapidly, but Lori La France's Marine Science class at the Ipswich High School is trying to figure it out.

Two students, Seneca Freylewe and Louie Hewitt, carefully weigh, measure and sex each crab, then test the color of the crabs' abdomens against a 12-point color chart from Wal-Mart. The abdomen colors change from green to red before the crabs shed their shells and mate.

The students have already developed the working hypothesis that both the male and female crabs move into the lower salinity upper reaches of the marsh in the spring but that as soon as the females start carrying eggs, they move back down toward the mouths of the estuaries where the waters are cool, salty, and hold more oxygen.

The students are also starting to talk about ways to control the crab population. The best way to do this would be to create a market for the predators. The town has already hired six fishermen to trap the crabs and sell them to Appleton Farms to be used as fertilizer at 25 cents a pound.

But Lori's Sustainability class has another idea. Most restaurants use chicken and fish stock for making chowders soups and seafood dishes, but do not because commercial crab stock is so expensive and difficult to find. But if the students could trap and boil the crabs down into a stock they could market it as something like:

Ipswich Soup Stock, Un produit de LaFrance.

I have already made some up myself and found it delicious. You can also just boil pasta in the stock to give it a savory, seafood accent. Knowing when the crabs are about to mate also makes it possible to serve the crabs soft-shelled.

The students are talking about using some of the cooked crabs as bait to catch more crustaceans and recycling the rest in the town's compost facility. The crushed up shells could be donated to chicken owners to enhance their hens' eggshell growth, though more work would have to be done to make sure the eggs don't start tasting too "crabby."

The best thing would be if the students could come up with a gourmet item that doesn't require fuel to be shipped in from distant locations.

Anything is possible by adding wits to water. It has always been the way to earn a good living in New England and hopefully it will continue to keep our area so uniquely sustainable.

CHAPTER 24
Road Trip!

The first thing most people think about when you mention energy is, "How much will it cost me to fill up my car? So when I was asked to write about fuel efficiency, I decided to start at the very top and called Tesla motors. Their Model S was about the hottest car on the market.

But I soon discovered that Tesla doesn't have traditional dealerships. They sell their cars out of malls, and their nearest store front mall shop was in Natick.

I have to admit I felt like a bit of a phony. I told the salesman right up front that their cars were way out of my league and I didn't want to buy one I just wanted to write about one. "No problem," said the salesperson. He sounded a bit like one of those computer wizards you find in an Apple store.

Road trip!

"I can get you an appointment for next week, but we also have an opening on Saturday." I wanted to start the story right away and didn't bother to think about what the Natick Mall would be like the Saturday after Thanksgiving, so I signed up instantly.

The salesman also happened to mention that I could bring two or three friends. In the spirit of full disclosure I have to admit that I am not really a car person. I was one of those kids in school who could identify a hundred species of insects but couldn't tell you the difference between a Ford and a Chevy. As long as my vehicle gets me from point A to point B I'm happy; without incident, even better.

I knew the test drive would be largely wasted on me, so I called up a friend who fixes BMW's for a living and another who owns a Prius. I figured that they would know what the salesman was talking about, and while they were all sitting up front talking amps, volts and torque I could be free to sit in the back waving at pretty girls. It seemed like the job I was most suited for on this, our middle-aged road trip.

But I also wanted to discuss our three different strategies for cutting down on carbon emissions. Let's call my friends Mike and Steve for the sake of anonymity. Steve bought his Prius eight years ago and figures that in two more years what he has saved on gas will outweigh what he had to pay to buy his car new. His is the most widely accepted way to both spur the economy and save gas.

Mike makes a living taking none too gas efficient BMW's and souping them up into high performance gas-guzzlers. But he also knows how to drive his car so precisely that he can stay in the sweet spot to save gas and knows how to glide to a halt rather than using his brakes.

My strategy is to buy a used Volvo for about ten thousand bucks and keep it on the road for ten years. This way I can end up only spending about $3,000 a year for transportation and still feel virtuous that some manufacturer isn't going to have to emit 20 more tons of carbon dioxide, just to make me a new car. I call this my New Englander strategy; you are free to call it my cheapskate strategy if you must.

We soon discovered what the Natick Mall was like on Thanksgiving weekend. But we also discovered what a stroke of genius it is to sell cars out of a mall. Thousands of gawkers were stopping by to take a look. Few of them were serious buyers, but they would all go home and rave about sitting in a brand new Tesla. Talk about free advertising.

The salesman, whom I shall call Vasilios, asked me if I wanted to drive a red or black model S. That was easy, the red one of course, the one that goes for about $80,000. Hmm, I could buy 8 used Volvos for that and drive them in succession for the next hundred years.

The model S doesn't have anything like fancy seagull wings, and to my untrained eye was actually fairly nondescript. I probably wouldn't be able to tell it from several other makes and models on the highway, but even I could tell that this was a spectacular piece of high-performance technology.

The exterior was so aerodynamically designed that there wasn't even a fuel cap to interrupt the car's flow lines. There was just a small circle of glowing green lights to show you where to put in the plug instead. This futuristic design also caused me my first problem. The exterior was so aerodynamic I couldn't find a door handle to get into the car. It was only when I happened to touch a piece of shiny ornamental chrome that a handle eased out of the side of the car as silently as something you might expect to find on the Star Ship Enterprise.

My second problem was that I decided to let Mike drive. I didn't want to be held responsible for dinging up $80,000 car. But that was before I heard Mike casually mention that he had formerly been a racecar driver. Hey, how bad could that be?

We drove out of the mall and Vasilios found us a narrow little rural road, where he then said, "OK, now here is where you can try to accelerate it a bit." I was thrown deep into my seat then projected forward and found myself floating in zero gravity as we came up rather too rapidly to the car in front of us.

I said something like, "Holy Moly Michael, how fast were you going?" Mike explained that racecar drivers are taught to never look at the speedometer, just to concentrate on driving. That was comforting!

The next time we came to a straightaway section I reversed Mike's strategy and concentrated on taking an incriminating photograph of the speedometer rather than looking at the world's narrowest rural road flecked with slippery patches of ice and snow.

I could already see the wheels were turning in Michael's head. Several years ago General Motors was working on a concept to design a median priced electric car but dropped the idea because they figured they wouldn't be able to convince energy companies to put in charging stations. So GM went into bankruptcy instead…good plan.

But Elon Musk, who started Tesla, figured that if he designed a really cool electric car and set up his own super charging stations, Tesla would be able to sell enough luxury cars to be able to gradually reduce the price of each car to about $40,000. He seemed to be on to something. Porsche, Audi, BMW and Mercedes all announced they would soon be coming out with all electric luxury cars. But none of their models would be available before 2017.

With its advanced lithium battery and 300 mile range you can already drive a Model S cross country, stopping about every six hours to recharge the car and enjoy a leisurely hour long meal at one of Tesla's 140 supercharge stations, set up in the last two years.

By 2017, Tesla will have built several hundred more supercharge stations and be well on its way to bringing down the price of a new model to where a guy like Mike can start thinking about buying one.

But the test drive also gave me a glimpse of the future. Several of the people who are now buying Teslas live in solar homes and work for companies that provide special charging outlets to any employee who has invested in an electric car.

When we returned home, Steve told mo ho had plans to sell the first floor of his three-story home as a condominium and put the money into installing photoelectric panels on his roof so can charge up an electric car.

But, I'm still a lost cause. I figure I can continue to eke another hundred thousand miles out of my old Volvo. By then, Mike might be ready to sell me his used Tesla.

CHAPTER 25
The Elf and the Water Maker;
Newburyport, Massachusetts

Elon Musk was worried, not because Tesla's stocks had dropped. He expected that. He was worried because Toyota had just announced it was going to start selling hydrogen fueled cars in Japan and in the United States by June.

Satoshi Ogiso was saying that electric cars were just a temporary stepping-stone on the road to a full hydrogen economy. The managing director of Toyota was so sure of himself that Toyota had scaled back its research on all electric cars. It was gambling that the future would be based on hybrid cars like Toyota's wildly successful Prius and their new hydrogen-fueled Mirai, which means future in Japanese.

Conrad Willeman in the Elf

Ogiso pointed out that you could fill a Mirai in five minutes, while it took an hour to recharge a Tesla. But Musk knew that his advantage lay in the 140 supercharging stations that were already providing free electricity to Tesla owners, and the 160 more that would be finished by 2016. That infrastructure was already built into his $50,000 base priced automobile.

Toyota had countered by offering three years of free hydrogen to Mirai's initial buyers. But the hydrogen would be hard to find. There were only 26 hydrogen refueling stations in California and only 46 were planned for 2016.

Hydrogen is the most abundant element in the universe, but it is difficult to produce. The most common way to make it is to combine methane with high temperature steam, which releases both hydrogen and a smaller amount of carbon dioxide, which can be removed through pressure-swing adsorption. This means that fracking for methane could continue to be important in a hydrogen economy. Eventually hydrogen will only cost about a dollar a gallon to produce. And when you burn it the only "pollutant" produced is water. After all, hydrogen means "water maker" in Greek.

But hydrogen can also be produced by using photo electro synthesis cells to strip hydrogen atoms off water molecules without producing any pollution. Toyota has a contract with Sun Hydro to open such a filling station soon in Connecticut and in Braintree, Massachusetts.

Hydrogen's biggest problem could be its image. What happens when a Mirai flips over and explodes in a ball of fire bringing back memories of the *Von Hindenburg* dirigible accident? Toyota claims that their fuel tanks meet all the industry's standards for holding hydrogen, but so did the *Von Hindenburg's*.

The other problem that Toyota had is that they wanted their new car to stand out. It does but in the wrong way. The Mirai looks like a cross between Darth Vader and a samurai swordsman coming at you, and a Prius going away.

The Mirai wasn't going to be available in the United States until June so I decided I would test drive the world's cutest vehicle instead. That would be the Elf made in Durham, North Carolina.

But it was just before Christmas and it seemed counterintuitive to fly to North Carolina to see an Elf, so I called Conrad Willeman who lives in Newburyport and owns an Elf. He suggested we meet on a sunny day on Marlboro Street.

I soon discovered the reason for the sunny day. The Elf is an enclosed three-wheeled bicycle with lights; directional signals and an electric motor to either assist it on hills or power it independently.

But the nicest thing about the Elf is that you don't have to worry about a racecar driver wanting to test it like a Tesla, or that it might explode in a ball of flames like a Mirai. You just sit back and drive it in the slow lane like a regular car or pedal it like a bicycle along with its electric motor assist. And if you live in Newburyport, which gets a fifth of its electricity from the True North Solar Farm you have a vehicle run on sunshine and sweat. How can you beat that?

Chapter 26
"Liken Him unto a Foolish Man Who Builds his House on Sand." -Mathew 7:24
Plum Island, Massachusetts

One of the best ways to limit the amount of carbon dioxide emitted into the atmosphere would be to stop building so many houses, but there are some problems with that solution.

The West likes to point out how much carbon dioxide China produces to build homes for their newly rich. China likes to point out that the West produces 56 tons of carbon dioxide to fell lumber and run the machinery to build an average American home, and it is more like 100 tons if the building is made out of concrete. Plus, how do you convince economists to stop using "new housing starts " as their indicator of a "healthy" economy? As if you can have a healthy economy without a healthy environment.

But one thing people in both the developed and the undeveloped world can do is stop building homes where nature will continue to destroy them. One of the best places to observe this curious behavior is our neighboring Plum Island. It has become New England's poster child for everything a community should not do to a barrier beach.

For years people built summer fishing shacks on the island. They were mostly made out of driftwood, had no electricity or running water and were worth about $3,000 or $4,000. If the shacks were threated by a storm the owners could simply move them back behind the dunes or walk away quickly without losing their shirts.

A house built on sand, Plum Island MA

But after about a century of such behavior, so many houses had been moved behind the dunes that the island was as crowded as most cities and its inhabitants started getting sick because their water wells were being contaminated by their neighbors' outhouses.

So in 2004, the EPA convinced the towns of Newbury and Newburyport to bend every conceivable state and local environmental regulation in order to bury water and sewer lines underneath their barrier beach. The cash strapped communities smiled as tax revenues rolled in, as the homeowners remodeled their fishing shacks into permanent year round homes, some worth up to a million dollars.

But a barrier beach needs to be able to move, pulse and migrate in order to stay healthy. Plus, the island was beginning to be slowly inundated by the rising ocean at the same it was time starting to run out of its reserve of sand, built up during the last Ice Age.

Eventually the inevitable happened and in 2008, houses started tumbling into the ocean in earnest. But by then, homeowners had sunk their entire life savings into their year-round beach houses, and the federal Flood insurance program would only reimburse them if they rebuilt their homes on exactly the same footprint as before!

So far the worst year has been 2013, when thirty-nine houses were declared uninhabitable, seven were eventually lost and one was wisely moved to the landward side of the street.

By this time the cash strapped town of Newbury was getting 40% of it's tax revenues from the houses on Plum Island so its Board of Selectmen declared a state of emergency and state officials looked the other way as homeowners built a half mile long, illegal seawall. Then they compounded the problem by encouraging even more people to build new homes in the dunes that were destined to collapse in five to ten years.

The beach was eroding just as fast as before, but now it was happening out of sight, behind and under the seawall. The erosion only became apparent a few days after a storm, when the ocean had sucked sand out from behind the seawall creating a vacuum. This caused cracks and fissures to appear in the sand dune behind the seawall until eventually the whole thing collapsed, carrying more houses away with it.

This process will keep happening until all the 40 homes condemned in 2013 have been lost again, in just a few short years. But each time one of these houses is washed away and rebuilt, 56 more tons of carbon dioxide is emitted back into the atmosphere, and the process speeds up even faster.

My neighbor's solar home

CHAPTER 27
My Solar Neighbor
Ipswich, MA

November 17th was probably not the best day to visit the first star certified solar house in Ipswich, Massachusetts. It was cold dark and rainy. The polar vortex was bearing down on New England, but the home's electric meter showed that the roof mounted photoelectric panels were still producing electricity, and that its thermal panels were still delivering warm water to heat the house.

The owner told me that Azimuth Construction had installed photoelectric cells specially designed to operate in overcast climates. "These cells do not produce as much electricity on a hot summer day as other cells, but they are more far efficient on cloudy days."

Inside, the house is dominated by an open staircase that winds around the chimney and up through an airy three-story atrium. The house was built three stories high, so the tall surrounding oak trees would not shade the panels.

A fan at the top of the stairs was quietly circulating hot air back down to the first floor. In the summer the fan draws heat out of the house, making air conditioning unnecessary.

Best of all, my bare feet were being warmed by radiant heat emanating from large terracotta colored tiles on the floors of the dining room, living room and kitchen. All the water collected in the roof top panels was being circulated down through the house and under the tiles to create this cozy heat.

The owner dragged me from the tiles to show me where the warm water enters the house in the attic. Then pipes carry it down into the basement where it is heated to 130 degrees by the solar powered hot water heater, then circulated back under all those warm, cozy floor tiles.

Most solar houses have heated swimming pools because the houses produce so much extra hot water. But these owners were New Englanders and decided to dispense with the pool and store the water in a cistern instead.

The panels are so efficient they can produce enough energy to heat the house, run 6 computers, a high-energy router and still have enough left over electricity to earn a little money back from the town's municipal power station on bright sunny days.

The house could have also produced enough electricity to charge an electric car, but such cars require a garage and the owners ran out of money. This was before the days of tax rebates and generous solar energy incentives. During the coldest days of the year their small gas burner provides ancillary heat and two wood burning stoves provide atmosphere.

But enough of all that technical stuff, I really wanted to go back down and pad around on all those nice warm tiles before returning to my own house that was feeling every bit as cold and drafty as its 144 years would suggest!

CHAPTER 28
Putin's Problem
The Kara Sea | September 26, 2014

Texas steaks coupled with Russian vodka and briny Beluga caviar topped the menu aboard the drill ship *West Alpha*. The president of Rostneft was toasting his combined Russian—American—Norwegian crew. They had just discovered more oil than resides below the entire U.S. section of the Gulf of Mexico.

But Igor Sechin's jubilation was tempered. He knew his Kara Sea project was officially dead. The United States had just put sanctions on his company because Russia had invaded the Ukraine. His good friend Putin knew all about politics but didn't know "zynayet der'mo" about business.

The head of Exxon Mobile, Rex Tillerson, was equally upset with President Obama. Didn't those politicians know anything about making money? Didn't they know that it was only the big oil companies who could see 50 years down the road, instead of just to the next election?

Sechin's problem was that Rosneft couldn't go it alone. They needed partners like Exxon Mobile, BP and Norway's Statoil that had the technology to drill in such cold, deep, offshore waters. They were also part of his grand strategy. He knew that his Siberian oil fields were running out of oil. If he wanted to stay in business, he had to team up with these western companies to develop the profitable Kara Sea oil field. It was pivotal to his long-range plan to have Russia surpass America as the world's largest supplier of gas and oil. Didn't Putin know that getting oil out of the Arctic was far more important than anything going on in the Ukraine?

Tillerson was in an equally tight spot. Exxon also needed to develop the Kara Sea field just to maintain its position as the world's largest producer of energy. Her production had fallen off in both 2012 and 2013 and was now essentially flat in 2014.

Those American frackers had driven the price of oil so low that it was too damn expensive to drill in such forbidding places. They had become everybody's "bolshoya problema".

Fracking had turned the entire energy industry on its head. The world was awash in inexpensive oil and OPEC was not about to cut production to prop up prices in Russia and the United States.

Iran, Iraq and Saudi Arabia were all trying to out drill each other so they would have enough money to fight proxy battles in Syria, Iraq, and wherever else ISIS might pop its head up next. It was bloody hard work making a buck in such a rapidly warming world.

Of course Igor didn't say any of these things when he stepped up to the microphones for his on-board press conference. Instead he smiled and simply said, "This is our united victory. It was achieved thanks to our friends and partners in Exxon Mobile, Nord Atlantic Drilling, Schlumberger and Halliburton. In honor of all of them we would like to name this field Pobeda, which means victory in Russian." That man must have some mighty good spin-doctors back in Moscow.

CHAPTER 29
Capitalism vs. the Climate?
New York City

"There is plenty of room to make a profit in a zero-carbon economy; but the profit motive is not going to be the mid-wife for that great transformation."

-Naomi Klein, 2014

On a beautiful September day marking the Autumn Equinox, half a million people in 166 countries protested against climate change. So many people marched in New York City alone that organizers had to tell them to disperse because they were clogging the streets between Central Park West and the United Nations building, where the U.N. conference on Climate Change would convene the following day.

40,000 people marched along the Thames in London and 25,000 people rode their bikes beside the Seine in Paris. The same was true in Australia, India, Tonga and Tuvalu, island nations that would soon be underwater.

Many of the protesters' signs reflected the fashionable theme that the way to get rid of climate change is to get rid of capitalism. But on the same day the Rockefeller Brother's Fund, if not a bastion of capitalism at least a capitalistic running dog, announced that it planned to join 180 churches, colleges and other philanthropic organizations in selling over $50 billion worth of investments in companies involved with fossil fuels.

Two days later, CEOs from many of the Fortune 500 companies sat down with Ban Ki-moon, the Secretary General of the UN, to outline what they intended to do to slow climate change. Kellogg, Unilever, Cargill and Nestle all pledged to stop deforesting to make way for Palm Oil plantations, which rob the earth of nature's ability to sequester carbon dioxide.

Apple's chief executive, Tim Cook pointed out that his company was already using renewable energy to power its data storage facilities in the United States, and it pledged to start having its suppliers in other countries switch to renewables as well.

But he categorically rejected the argument that society must choose between capitalism and the environment, pointing out that Apple built a major solar farm in North Carolina to help power their data center in that state, and has required that power companies in states like Iowa would have to start purchasing significant amounts of renewable energy to have Apple consider locating its facilities there as well.

The head of Greenpeace International, Kumi Naidoo, pointed out that the world could not just count on the goodwill of such corporate leaders, but that government officials also had to continue taking action through the Conference.

These gestures of protesters, philanthropists, executives and diplomats were largely symbolic, but at the same time, we have seen that the world is in the early stages of what could become a rapid transformation of the energy field.

However, this transformation was not being driven by the radical activists, lofty philanthropists, exalted executives and political leaders represented in and outside the conference, so much as by third generation hot water manufacturers like Jim Vaughn who built the largest solar farm in New England, stubborn businessmen like Jim Gordon who bucked drawn-out opposition in his efforts to build America's first offshore windfarm, even by rough-hewn wildcatters like George Mitchell who started the fracking boom.

These are the people who have driven down the price of land based oil and gas so much that large oil companies have started cutting back on their expensive efforts to explore and drill for offshore oil. These are the people who have driven down the price of renewable energy so much that traditional power companies are now either switching fuels or going out of business to be replaced by smaller, cleaner, less centralized renewable energy facilities. They are also the people who have helped countries like Denmark, Germany, China and Korea manufacture cheaper wind turbines and solar cells to bring about this transformation.

Regulations, incentives and government subsidies have all helped, but in the end it was people making bold use of capital who made these things happen, and these changes are happening quietly but rapidly all over the world.

In fact almost all of the world's rapid transformations have been made possible by focusing the immense force of capitalism. Capitalism is merely a means to accomplish an end. But it must be steered, harnessed and filtered through the laws, regulations and values of humans to produce the ends we want to accomplish. Let's focus on what these ends are, rather than wasting time fighting against one of the most effective tools in our arsenal to slow climate change. Let's not throw out the baby with the bathwater.

CHAPTER 30
Think Global, Act Local
October 17, 2014

The Climate Change protests in New York and other cities around the world highlighted the fact that was a groundswell of broad-based concern about global warming, but it also put the environmental community on notice that members of environmental organizations were sick of having their leaders coopted by the fossil fuel industry.

When the modern ecology movement first got underway in the Seventies, bipartisan coalitions of scientists and congressmen worked beside environmental organizations to craft twenty-seven pieces of far reaching environmental legislation signed into law by none other than Richard Millhouse Nixon. Not because he was necessarily in favor of the environment mind you, but because the time was right. During that first wave of awareness, environmentalists pushed for straightforward limits on pollutants and their actions helped stop acid rain and shrank the ozone hole.

But then, along came the genial anti-environmentalism of Ronald Reagan and James Watt. Unrestrained capitalism and deregulation become the zeitgeist of the Eighties and Nineties and leaders of many of the larger, slicker environmental organizations leapt into bed with the fossil fuel industries.

The heads of the Sierra Club and the Nature Conservancy could be seen flying off on corporate jets to make deals and form partnerships with the likes of Exxon Mobil and Shell Oil. The Nature Conservancy not only allowed Mobil to continue drilling oil on land they had given to the Conservancy to protect one of the last remaining populations of prairie hens but the Conservancy then went on to start drilling its own well to fill up their non-profit coffers.

The peak of this cooptation came about when the major environmental groups teamed up with their oil executive cronies to craft a cap and trade plan for President Obama to present to Congress. The fossil fuel companies played the environmentalists like a violin, making sure the legislation had no chance of passing. Obama had to cut his losses and turn his attention to health insurance reform and saving the economy that the Bush administration had left on palliative care only.

It was only in his last two years in office that Obama realized that he had the power to do things like form a broad-based coalition to defeat ISIS and direct the EPA to force companies to lower carbon emissions enough so that the companies started to switch to natural gas and renewable energy, which were not only cleaner, but also starting to come down in price.

The marches also made clear that the old way of wheeling and dealing with the fossil fuel industry was outdated, and if the larger environmental organizations did not pay attention they stood to be replaced by a network of new, more focused local organizations willing to take on polluters and anti-environmentalists in a far more straightforward manner.

CHAPTER 31
Putin's Dilemma
Brisbane, Australia | November 16, 2014

President Obama strode to the podium, paused dramatically, and then turned to Vladimir Putin saying, "After the mid-term elections I'm delighted to see a friendly face."

Unfortunately his clever barb never happened at the G-20 meeting in Brisbane, just in Conan O'Brien's opening monologue in downtown New York City. Brisbane was a far chillier scene, with baleful stares and frosty exchanges.

Putin really didn't like President Obama. Here he was, almost a grandmaster at chess, playing the age-old game of international politics, and there was that lightweight with his goofy grin. Putin knew how to use coercion and realpolitik to control the board. His end game was to dominate Eurasia until he had surpassed the United States as the premier power on earth.

The problem was, he was losing. Those fracking American drillers were pumping so much oil that the Russian economy was on the verge of collapse. The ruble had already plunged 23 percent and Russia's Central Bank was forecasting zero growth for 2015.

Putin had to swallow a recent deal to sell gas to China for far less than he had become used to selling it to Europe, and the those damn sanctions had torpedoed his contract with companies like Exxon to search for oil in the Arctic Ocean and Siberia.

Poland, and the former Soviet republics of Latvia, Lithuania and Estonia had become stalwart members of NATO, why even Sweden and Finland were thinking of joining!

And then there was the Ukraine. Putin's annexation of Crimea had strengthened Europe's determination to get over its addiction to cheap Russian oil.

But Putin's main problem was that his American counterpart didn't even know how to play chess! Obama was playing a sophisticated game of Monopoly, where he was the Capitalist banker slowly squeezing the life out of the Russian economy. At the same time, Obama was strengthening his position as head of the Atlantic Alliance. His moves might not elicit shock and awe but his sanctions were pretty darn awesome.

It had all come to a head in Australia, still smarting from the loss of its 38 nationals killed when the Malaysian airlines had been shot down over the Ukraine. The prime minister had even tried to exclude Putin from attending the conference. But cooler heads prevailed and Putin had faced a withering succession of Western leaders browbeating him for supporting the Ukrainian secessionists.

The prime minister of Canada, Stephen Harper, only reluctantly shook Putin's outstretched hand saying, "I will shake your hand but I have only one thing to say to you. You need to get out of the Ukraine." But behind the scenes Putin was letting it be known that in fact he would pull back from the Ukraine as long as he could be sure that NATO would not put tanks and missiles on his vulnerable western border.

By the end of what the Australian prime minister famously called these "shirt-fronting sessions", Putin had tucked his tail between his legs and flown back to Moscow saying that he needed to leave early before starting work Monday morning. Nobody believed him.

Even though his bare chested anti-American nationalism still had legs, Putin was flying back to a Russia where people were growing restive from the bite of the Western sanctions.

Obama, on the other hand, was flying back to a country where the economy was recovering, unemployment was in decline and drivers were paying $2.80 for gas. Yet his approval rating was at an all-time low, probably from not enough chest thumping. Go figure.

CHAPTER 32
OPEC
Vienna

Ali Al-Naimi was getting sick and tired of those scrappy little Texas frackers. Who did the illiterate infidels think they were? The Saudi Arabian oil minister held degrees from Lehigh and Stanford and was on a first name basis with the heads of all the Western oil companies.

But Ali had a plan to teach the frackers a lesson. Just flood the market with cheap oil and wait for the prices to drop. All he had to do was convince the other OPEC nations to go along.

Venezuela, Ecuador and Libya all wanted to cut production to keep prices high. Iran did too, but made it known that "under no circumstances" would it cut even "a single barrel of its own oil" from production. It had already sacrificed a million barrels a day because of sanctions, and its position was growing steadily weaker in its negotiations with the West over its nuclear program.

But Ali knew he held all the cards. Saudi Arabia was the only country that could keep pumping oil out of the ground at $30 a barrel and still make money. The other OPEC countries, and even Russia, needed the price to stay at about $100 a barrel to keep from running deficits.

Ali didn't have to worry about deficits but he did have worry about the 9 million barrels of oil the frackers were pumping out of America's shale deposits every day. That was the real threat to his position as de facto head of OPEC. It was much more important to teach those frackers a lesson than to sidle up to Iran who was forging an unholy alliance with Iraq to fight ISIS anyway.

The United States had always been Saudi Arabia's staunchest ally in the Middle East because of America's dependence on Saudi oil. The frackers had changed all that, but if Ali could force oil down to $75 a barrel it would put the frackers out of business in West Texas, Oklahoma, Louisiana and Arkansas. Still that only accounted for about 400,000 barrels a day. If he could force the price of oil below $70 he could also put the frackers out of business in North Dakota and East Texas. Those two regions alone accounted for more than half of America's gas and oil.

Ali's plan was to cook the frackers turkey on Thanksgiving Day. That was when the OPEC countries held their annual meeting in Vienna, and he was easily able to convince them to keep the production of oil at 30 million barrels a day.

He arranged to have the decision announced at just about the same time the frackers were sitting down to watch the Philadelphia Eagles pommel the Dallas cowboys. He remembered the Eagles from his days at Lehigh. That would give the frackers something to think about while they were digesting their Thanksgiving Day turkey.

CHAPTER 33
Obama's Legacy
Washington D.C.

"Go ahead make my day, " smiled Obama to himself. His Republican enemies had warned him not to reform the country's immigration policy but here he was stopping the deportation of over 5 million hard working immigrants from our nation of hard working immigrants.

The 44th president was way too cool to taunt his enemies like Ronald Reagan. But he was also too smart to be cowed. Immigration reform was the right thing to do and he knew he had authority to do it.

But even Obama was surprised by the response. You could almost hear the jubilation rising out of Spanish Harlem. Univision had delayed broadcast of the Latino Grammy awards just to carry the president's full address. Thousands of Mexicans were busily printing photographs of the president Obama to hang on their walls, right beside the Virgin of Guadalupe.

Republicans suddenly realized they were now dealing with a newly rejuvenated president. The mid-term elections were over and Obama no longer had to worry about being reelected. He was free to solve the problems he had put on hold in 2009 in order to deal with the recession left in haste by President Bush.

However, Obama's greatest success was in the environmental field. He had already accomplished more than any other president since Theodore Roosevelt, but he had done it without passing a single law.

The President had accomplished all this with his characteristic care and meticulous legal planning. Remember, he had graduated Magna Cum Laude from Harvard Law School and gone on to teach constitutional law at the University of Chicago. He was probably the best-trained legal scholar to ever sit in the Oval Office. He certainly knew his Constitutional powers and the system of checks and balances between his office and the Supreme Court.

After Republican senators thwarted his efforts to push a climate change law through Congress in the beginning of his first term, Obama had directed his team of crack legal scholars to turn their attention to the Clean Air Act. It had been passed unanimously and signed into law by President Nixon. It was one of the most comprehensive bills ever passed, but no president had ever used it to its full capacity.

George Bush Senior had updated the bill to regulate ozone and mercury, but Obama was the first president to use it to fight global warming. After his cap and trade bill failed to pass, he had the EPA promulgate regulations that required all cars get at least 55 miles to the gallon by 2025.

The regulations led to the rapid development of hybrid and electric cars and had convinced Elon Musk to start building his Teslas. When the grid finally switches over entirely to renewable energy, such electric and hydrogen fueled cars will reduce America's carbon emissions by close to a third.

In his second term, Obama ordered the EPA to crack down on power plant emissions. This made the electrical industry start switching to cleaner fuels which had allowed Obama to fly to China and make a joint announcement with President Jinping that the top two polluters in the world had signed an accord to cut their emissions by close to a third over the next two decades, not the whole solution but a significant step in the right direction.

By the end of 2014 Obama would also direct the EPA to regulate methane emissions. This would curb the fracking industry, where regulation was so sorely needed. The regulations were timed so they would be in place by the time the frackers had recovered from OPEC's actions and were ready to drill again.

In six short years Obama had set in motion a transformation in the energy field that might be enough to allow our planet to avoid the worst of the so called runaway Venus scenario that Carl Sagan warned of back in the 1970's. It would cap Obama's legacy as one of the best environmental presidents ever elected in the United States, and in the world, if he could convince leaders of enough nations to pass a comprehensive climate treaty at the conference scheduled in Paris in late 2015. Now that would be a legacy to be proud of.

CHAPTER 34
Dead Rats
Lima, Peru

"There are dead rats that need to be swallowed," said the diplomat from New Zealand." "If you are submitting for circumcision be very careful that it doesn't become an amputation," said her colleague from Singapore. The male delegates winced at such undiplomatic language.

But why were the delegates at the 2014 climate conference so upset? After two weeks of talks that ran several days over the time half the delegates needed to get home for Christmas break, they had finally hammered out a draft agreement to be ratified in Paris by late 2015. Of course the French were already calling the yet to be signed deal the "Paris Alliance."

Obama had made this "Lima call for climate change" possible by ordering the EPA to decrease carbon emissions at electrical plants and increase automobile mileage by 2025.

It was the first time the United States had ever come to one of these meetings with an actual domestic carbon policy in hand. We had usually just demanded that other countries cut their use of coal and carbon emissions while only offering vague promises that we would do the same, sometime in the far distant future.

Our main rival at these conferences had always been China, insisting that developing countries should not have to make any cuts because they had to catch up economically first. Obama had broken through this impasse by wooing President Xi Jinping to jointly agree to reduce emissions predicated on his EPA actions. This had broken the logjam leading to the new draft to be signed, hopefully, in 2015.

But the 196 delegations still had to overcome some significant hurdles. Every country was supposed to come up with a detailed plan for how it intended to cut emissions before March. India had already said that it would not be able to have a plan before June. So any other late announcers would also have to receive a pass. But the point was to gather enough information to finalize the Paris Agreement, not to be punitive.

It was pretty crazy to allow each country to base their reductions on their own political situation, but it was also the only way to get an agreement. And of course there were always wild cards. Australia's emissions had soared after it repealed its recent climate change law, and the United States delegation was dealing with its "Republican problem" by urging that the final agreement not be a legally binding treaty so it would not have to be voted on by the Senate.

Of course the disagreements would probably all resurface at the Paris Conference, but the delegates could go home now, knowing they had at least reached a draft agreement. The kicker was, at its very best, the agreement would only reduce half the emissions necessary to avoid the worst scenarios for sea level rise, drought, and violent storms. This was only a small step for man and an even smaller step for mankind, but after 17 years of negotiations who was counting? At least it was a start.

CHAPTER 35
Big Oil; The Beginning of the End?

Premier Aleqa Hammond put down the phone. How was she going to explain this to her constituents? The CEO of Statoil had just informed her that they were handing back their licenses to explore for oil off the west coast of Greenland.

Greenlanders had been counting on their offshore oil to supply income so they could afford to seek independence from Denmark. Aleqa had been promising her Inuit constituents that Greenland would become the first nation of indigenous people in their lifetime.

But the same thing was happening all over the world. Royal Dutch Shell had just abandoned plans to build a $6.5 billion petrochemical plant in Qatar. UK's Premium Oil had announced it was delaying a decision on whether to go ahead with Project Sea Lion, its proposed deep-sea drilling operation off the Falkland Islands.

The entire industry was shattering; workers were being laid off, supplies canceled, drill ships were lying idle in desolate harbors. Oil companies were pulling out of Africa, the Arctic, and the deep sea off Brazil.

The big story at the end of 2014 was that the price of oil had plummeted to $46 a barrel, half of what it had sold for in June. But the big story at the beginning of 2015 was the rapid collapse of the oil industry itself. Even Lord Browne, the former head of BP, said that oil companies were in denial about climate change because, "They could not acknowledge an existential threat to their business."

The majors used to pride themselves on being bigger, more powerful and far thinking than any single country. They had their own ships, their own banks, guards and intelligence units. If all else failed they had always been able to rely on the armed forces of the United States, Great Britain, even Russia to protect their monopolies in Iran, Iraq and Saudi Arabia.

They harkened back to the days of John D. Rockefeller and the head of Occidental Oil, Armand Hammer who was the go between for 5 General Secretaries of the Soviet Union and 7 presidents of the United States. And of course Exxon and BP were lobbying to let Vladimir Putin off the hook. Who was going to let a small thing like the invasion of Crimea get in the way of drilling for oil in the ice free Arctic Ocean?

The companies had snickered at environmentalists' attempts to get schools and churches to divest themselves of oil stocks. They had sneered at frackers penny ante attempts to get residual oil out of old worn out wells.

It was only when OPEC decided not to cut back on production that they started to realize that they could be left holding a bag of stranded assets — over a billion dollars worth of oil in fields that could only be profitably drilled if the price of oil was over a hundred dollars.

Big Oil simply couldn't compete if oil was selling below $50 a barrel, demand was dropping as countries started limiting carbon emissions and hard-nosed investors were starting to realize the problem of stranded assets wasn't going to simply disappear. Over $91 billion dollars of their investments were at risk.

Everyone knew that oil would eventually rebound, but the big oilmen also knew that the damn frackers could get back into the game faster than they ever could. The majors had been running their companies like Captain Ahab obsessed with capturing Moby Dick. They had built bigger and more expensive drill ships to ply further and further afield to drill for more expensive oil.

They had ignored the threat of frackers like 19th century whalers had ignored Edwin Drake, the faux Colonel who had secured title over land in Hicksville Pennsylvania and was busy drilling for rock oil and selling it out of discarded whiskey barrels for far less than a barrel of whale oil.

The writing was on the wall, but like their 18th century counterparts, the modern oilmen couldn't believe what they were reading.

CHAPTER 36
Another Rockefeller?

"It turns out there are times when a paranoid control freak is just what the occasion calls for."

-Calvin Trillin, "Tepper Isn't Going Out; A book About Parking in New York." 2002

On January 14th the EPA promulgated new rules for regulating methane emissions and a Congressman or two argued that 2015 would be a good time to pass a carbon tax, because the price of oil was so low nobody would notice.

Meanwhile, environmentalists were still protesting against the Kinder— Morgan pipeline in Massachusetts because it would bring natural gas to the state from the fracking fields of Pennsylvania. The problem was, Massachusetts needed that natural gas. No matter how many wind and solar farms were built, the state would still need plants to generate power when the sun wasn't shining and the wind wasn't blowing.

It was ironic that the focus of concern was on the fracking practices in Pennsylvania. The last time Massachusetts was concerned about Pennsylvania was when Edwin Drake started drilling for oil in 1859.

Ten years later Pennsylvania's foothills were cheek by jowl in oil derricks. Farms that had been worthless a month before were being bought for a million dollars one week, and sold for two million the next. It was a chaotic scene of boomtowns and ghost towns, corruption, waste, and muddy oil.

The principle businesses of towns like Pithole, Pennsylvania were liquor and leases. Hundreds of wildcatters were so eager to get rich quick that they would flush the wells to get the oil out of the ground as quickly as possible. This would exhaust the gas pressure and ruin the oil field. In the words of John D. Rockefeller, "Butchers, bakers and candle stick makers were all starting their own refineries and making substandard oil."

But then prices plummeted to $2.40 a barrel. Many wildcatters stopped drilling but others did not, driving the prices even lower.

During the flush days, oil was transported by horse drawn wagons. Hundreds of them would become mired in the pervasive mud. But since they were the only game in town, the teamsters charged exorbitant prices. It cost more to carry a barrel of oil to a nearby railroad station than to transport it by rail from Pithole to New York. Gradually however, oilmen started building their own wooden pipelines. It was a much cheaper way to transport the oil and created far less waste and spillage.

All this chaos and disorder disgusted a tall, aesthetic, detail oriented man named John D. Rockefeller; today he would be called, like Mayor Giulliani, a paranoid control freak. But there are times when you need a control freak and 1865, like New York in 2001, was just such a time. Tall, thin and unsettlingly quiet, this pious businessman set about buying oil fields, controlling the railways, and building his own refineries that produced "Standard Oil."

Rockefeller made the oil business safe and profitable by giving his competitors a "good sweating" until they sold their refineries to him. All his business communications were carried out in code. His name for Standard Oil, and perhaps himself, was "Morose."

But by 1879 Standard Oil owned 90% of America's refinery capacity, 90% of its pipelines and of course Rockefeller became known as a ruthless monopolist. But unlike the other robber barons of his time that had made their fortunes through speculation and money manipulation, Rockefeller had taken a wildly unpredictable business and transformed it into a clean efficient industry. He had almost singlehandedly created his own energy transformation.

We can certainly hope that the present attempts to regulate the fracking industry are successful, but my money rests on the hope that somewhere out there either stomping through the dirty fracking fields or perhaps sitting in a quiet board room of a big oil company, there is another control freak equally repelled by the waste in the fracking world, who sees the potential to make natural gas the clean, efficient bridge we still desperately need get us to a future based on renewables and hydrogen energy. If there is, he or she will make a lot of money but we will all benefit.

CHAPTER 37
Showdown
Washington D.C.

Normally, presidents like to keep their State of the Union speeches under wraps until January 20th. But Barack Obama was feeling feisty.

He wanted to fire a shot across the Republicans' bow. He ordered his press secretary, Josh Earnest, to drop the bomb the same day the new Republican Congress was being sworn into office. Josh told the press, earnestly of course, that he doubted the President would sign any Keystone Pipeline legislation, because it would undermine the State Department's review of the project as well as prejudice a Nebraska court case that was challenging the pipeline's route.

But everyone knew that the Republican led House and Senate were both going to pass the bill, then Obama would veto it and the Democrats would still have enough votes to sustain his veto. However, the politicians would have all had their chance to strut and bellow in front of Fox cable news. It seemed like the beginning of another political charade.

The real reason that the bill would not pass, however, was that OPEC and the frackers had made the pipeline obsolete. The world was wallowing in oil. Who wanted to pay top dollar for the dirtiest, most expensive fuel on the market? Why not buy clean, cheap oil from the Middle East and leave the Tar Sands in Alberta's fertile soil where they belonged?

While everyone's attention was on the upcoming Senate vote, however, Republican congressmen were up to their own shenanigans. They passed a bill making it illegal for peer-reviewed scientists to advise the EPA on regulations because they didn't have a financial stake in the outcome! They argued without a trace of irony that this would make the process more transparent.

"I get it," said Congressman McGovern from Massachusetts, "You don't like scientists and you don't want science to interfere with your corporate clients."

The White House simply said the President would veto the bill. It all made for some pretty good political theater, but a much more serious development was brewing up north.

On January 7th Massachusetts' two largest electrical suppliers backed out of their agreements to purchase electricity from Cape Wind's proposed offshore wind farm. Their "out" was that Cape Wind had not met a December 1st deadline to obtain financing to build the offshore turbines.

Wealthy homeowners had delayed the project for so long that cheaper forms of renewable energy like solar and land-based wind power were now available, and coal fired plants were busily converting to cheaper natural gas. The electric companies didn't want to be stuck with buying expensive electricity from offshore wind facilities. It looked like it spelled doom for America's first offshore wind farm. Had OPEC and the frackers done it again?

CHAPTER 38
Teaching An Old Dog New Tricks.

On April 17, 2015, I attended a week of protests held to convince Harvard University to divest itself of its fossil fuel stocks. It was an eye-opening experience.

In 2012, I had helped establish a non-profit organization to aid coastal communities in dealing with the effects of sea level rise. For years I had been urging our members to stay narrowly focused on adapting to climate change and let the bigger national organizations deal with reducing global warming.

But I had a personal reason for my arguments as well. I had watched brilliant people spend their entire careers on achieving a global treaty to protect the oceans, only to have all their work wasted when Ronald Reagan refused to sign the Law of the Seas Treaty. After watching that debacle I had decided I would rather spend my time working on smaller local problems where you had a greater chance of making a real difference.

And quite frankly I also felt it was simply too late to stop climate change, and there was nothing that we could really do to fight the global economic machine that was hell bent on wresting every last natural resource out of the earth and converting it to cold hard cash. Our species seemed to be hard-wired into having more and more babies, and driving more and more gas guzzling cars.

January 20, 2015

But here was this group of bright, eager, sleep deprived young under-graduates free of such skepticism and willing to take on both Big Oil and the richest University in the world in one fell swoop. They were full of energy despite being in the midst of exams, having colds and having to finish writing their theses. God did it make me feel old!

By ignoring the odds of defeat they had already been successful. They had garnered the attention of the New York Times, Bloomberg News and the Wall Street Journal among many other news organizations. They had done it by making the point that the Rockefeller Foundation had decided to divest itself of all its fossil fuel holdings, after spending years trying to reform their unruly family cash cow, Exxon. Stanford was also way ahead of Harvard having already sold its stocks in coal companies, why even that bastion of radicalism, the World Bank had just announced they were in favor of divestment.

The students had also driven Harvard into the uncomfortable position of lamely defending the many wonderful things it had done to stop global warming. Its favorite seemed to be that one of its faculty members had developed an artificial leaf that mimicked photosynthesis.

The president of the university, Drew Faust, looked especially uncomfortable. She had skipped her midterms as an undergraduate in order to march in Selma, Alabama. She had to admit that Harvard had already divested itself of tobacco and companies that had been doing business in South Africa. Now she was trying to make the argument that Harvard had a responsibility to earn the highest return on its endowment, even though the University had lost $21 million dollars on it's fossil fuel stocks in just the last six months.

She had finally been reduced to arguing that divesting would be mean-ingless because other investors would simply snap up the stocks and make a killing. But, in the face of the rapid destruction of the planet, which the Corporation didn't dispute, did that really matter?

Perhaps the former President of Harvard, Derek Bok, said it best when the university had been deciding what to do about apartheid, "There are rare occasions when the very nature of a company's business makes it inappropriate for a university to invest in the enterprise."

It certainly seems like this is one of those occasions. I hate to encourage Harvard's already hyper inflated sense of itself but it is clear that the symbolic significance of convincing Harvard to divest itself would help trigger a cascade of similar actions, and that doing nothing would be an even bigger political statement.

But I took home another lesson as well. If these kids could take on Harvard and win, perhaps even this old dog could learn some new tricks!

FOOTNOTES:

Many of these chapters originally appeared as articles in the Newburyport Current. Chapter one, "The Accidental Oilman" originally appeared in the Christian Science Monitor Magazine and was serialized in Reader's Digest.

INTRODUCTION
Applebome, Peter. *They Used to Say Whale Oil was Indispensible Too.* New York Times, August 3, 2008.

CHAPTER 1 - AN ACCIDENTAL OILMAN
Charlie Gibson personal interview, Cappiello, Dina To Clean up Coal Obama PushesMore Oil Production. A.P. December 22, 2014.

Hefner, Robin. *The United States of Gas.* Foreign Affairs, May/June 2014.

McCarthy, Dan. *Fracking; The Good, the Bad and the Possibilities.* Fast Company 5/19 2014.

CHAPTER 2 - UNANIMOUS CONSENT
Hofmeister, John. *Why We Hate the Oil Companies.* Palgrave/MacMillan NY 2010.

CHAPTER 3 - BAKED ALASKA
Rosen, Yareth. *Noble Discoverer Leaves Port.* Alaska Dispatch News, August 28, 2014.

CHAPTER 4 - PUTIN'S PLOY
Kramer, Andrew. *Exxon Mobil moves Ahead With Russian Oil Drilling Project.* NY Times July 22, 2014.

Satell, Greg. *Putin Plays Chess. Obama Does Not...And that's why he's winning.* Forbes magazine July 1, 2014.

CHAPTER 5 - OIL AND EBOLA LAGOS
Ebola Delays Exxon's plans for Liberia. Petroleum Africa.com October 6, 2014.

Landers, Jim. *Ebola Stalls Mercy Ships,* Dallan News October 3, 2014

CHAPTER 6 - COOKED ALASKA
Rosen, Yereth. *Shell's New Chukchi Plan. Two Rigs Drilling Wells and the Same Time.* Alaska Dispatch News August 28, 2014.

CHAPTER 7 - CHAPTER 7 - STOP DEMONIZING NATURAL GAS
Lewis, Tanya. *Cause of Mysterious Siberian Hole Possibly Found.* Scientific American July 31, 2014.

Fountain, Henry. *Corralling Carbon before it Belches From Stack.* New York Times July 21, 2014

Krauss, Clifford. *Applying Creativity to a By-product of Oil Drilling.* New York Times December 17, 2013.

CHAPTER 8 - THE PIPELINE
Abel, David. *Protestors Voices Uneven Over Pipeline.* Boston Globe July 31, 2014.

CHAPTER 9 - THE VOYAGE OF THE CETACEA
Baker, Billy. *Three-hour whale Watch Lurches into Overnight Ordeal.* Boston Globe July 29, 2014.
Rosen, Andy. *The Whale Watch Boat Hit an Offshore Natural Gas Port.* What's That? Boston Globe July 29, 2014.

CHAPTER 10 - NUKES AND SEA LEVEL RISE
Evans-Brown, Sam. *Rising Tides in Seabrook. Is Nuclear Station Ready for Higher Seas?* Listen Line, New Hampshire Public Radio. April 9, 2013.

CHAPTER 11 - WIND TURBINES
Sargent, William. *The View From Strawberry Hill.* Strawberry Hill Press, Ipswich, MA, 2013.
Tuoti, Gerry. *Solar Power Increasing in Massachusetts.* Ipswich Chronicle June 14, 2014

CHAPTER 12 - WIND ENERGY IN THE GRAND MANNER
Selcraig, Bruce. *The Mayor of Wind. Sierra Club Magazine.* July/August 2014.

CHAPTER 13 - "TIS AN ILL WIND THAT BLOWS NO GOOD."
Gilles, Justin. *Sun and Wind Alter Global Landscape, Leaving Utilities Behind.* NYT Septmeber 13, 2014.

CHAPTER 14 - BORREGO SOLAR
Solis, Jennifer. *Controversial Solar Plan Passes Hurdle.* Newburyport News August 25, 2014.

CHAPTER 15 - TRUE NORTH
Norblett, Jackie. *Mass Eases growth of Solar Farms.* Boston Business Journal September 18, 2014

CHAPTER 16 - HYDROELECTRIC ENERGY
Ailsworth, Erin. *Legislation Raises Question. What is Clean Energy?* Boston Globe February 26, 2014.

CHAPTER 17 - GEOTHERMAL ENERGY
Rack Eric. *Magma From Iceland's Bardabunga Volcano Moving as Quakes Increasing.* Forbes. com August 26, 2014.
Mins, Christopher. *One Hot Island; Iceland's Renewable Geothermal Energy.* Scientific American August 20, 2008.

Chapter 18 - Locally Grown Geothermal

Shaffer, Pietro. *Celebrating the Power of the Sun; Creating the State's First Net-zero Energy Building.*

Chapter 19 - Biomass

Scheck, Justin. *Mass Tightens Rules on Biomass Plants.* Wall Street Journal August 21, 2013.
Arden, D.J. *Bermuda's Tynes Bay Waste to Energy Facility.* VirtualLibrary.com. 1991.

Chapter 20 - The Problem is Corn Not Drought

Sargent, William. *The Hottest Year on Record; The View From Strawberry Hill.* Strawberry Hill Press, Ipswich, MA. 2013.

Chapter 21 - Cows and Chickens

Sargent, William. *The Hottest year on Record; The View From Strawberry Hill.* Strawberry Hill Press, Ipswich, MA. 2013.

Chapter 22 - Black Gold

Our Company. Brickendsfarm.com.

Chapter 23 - Wits and Water

Sargent, William. *Wits and Water.* Newburyport Current.

Chapter 24 - Road Trip!

Rusick, Paul. *Tesla Faces Challenges from Mercedes. Porsche, Audi.* NY Daily News, November 26, 2014.

Chapter 25 - The Elf and the Water Maker.

Wysacki, Ken. *The Car that Runs on Sunshine and Sweat.* BBC June 9, 2014.
Flurry of Hydrogen Fuel cars Challenge all-electric Vehicles. Computer World, December 7,

Chapter 26 - "Liken Him Unto a Foolish Man."

Gillis, John R. *Why Sand is Disappearing.* OPED Section. New York Times, November 5, 2014.

Chapter 27 - My Solar Neighbor

In Balance; An Energy Efficient House. North Shore Life, Fall 2011.

Chapter 28 - Putin's Problem

Crooks, Ed. *Russia Curbs Put Exxon's Arctic Goals on Ice.* Financial Times. October 1, 2014.

Chapter 29- Capitalism vs. Climate

Schwartz, John Rockefellers. *Heirs to an Oil Fortune, Will Divest Charity of Fossil Fuels.* New York Times, September 21 2014.

MacComber, John. *The ABC's of Addressing Climate Change.* Forbes.com, September 24, 2014.

CHAPTER 30- THINK GLOBAL ACT LOCAL

Gillis, Justin. *Passing the Baton in Climate Change Effort.* New York Times, September 23 2014.

CHAPTER 31- PUTIN'S DILEMMA

Satell, Greg. *Putin Plays Chess. Obama Does Not...And that's why he's winning.* Forbes Magazine, July 1, 2014.

CHAPTER 32- OPEC

Lawler, Alex. *Inside OPEC Room, Naimi Declares Price War on U.S. Shale Oil.* Reuters, November 28 2014.

CHAPTER 33- OBAMA'S LEGACY

Eilperin, Juliet and Marken, Jerry. *Why Obama's Executive Action on Immigration Excluded Parents of "Dreamers."* Washington Post, November 30 2014.

Davenport, Coral. *In Climate Deal with China, Obama May Set 2016 Theme.* New York Times, November 12 2014.

CHAPTER 34- "DEAD RATS"

Priest, Marcus. *Lima Climate Conference May Bring Storm Clouds to Paris.* The Age, 12/14/14.

CHAPTER 35- BIG OIL: THE BEGINNING OF THE END?

Lyberth, Juaka. *A Greenlandic State or an Independent Greenland?* The Arctic Journal. June 3, 2014.

CHAPTER 36- ANOTHER ROCKEFELLER?

Klein, Naomi. *The Fracking Bridge is Already Burning.* Timechangeseverything.org. January 25, 2015.

Yergin, Daniel. *The Prize.* Simon and Schuster, NY, 1991.

CHAPTER 37- SHOWDOWN

Cappiello, Dina. *White house Says it will Veto Bill to Approve Oil Pipeline.* ABC News, January 6, 2015.

CHAPTER 38- TEACHING OLD DOG NEW TRICKS.

Howard, Emma. *Harvard Divestment Campaigners Gear Up For a Week of Action.* Guardian, April 13, 2015.

Harvard Faculty for Divestment Open Letter April 15, 2015.

www.ingramcontent.com/pod-product-compliance
Lightning Source LLC
Chambersburg PA
CBHW020836210326
41598CB00019B/1923